PERCEPTIONAL CARTOGRAPHY AND DISTRICT PLANNING

About the Author

N. D. Mani (*b.* 1956) took M.Sc. degree in Geography from Presidency College, Madras University in 1979 and Ph.D. in Perceptional Cartography from Gandhigram Rural Institute (Deemed University) in 1994. He is now Reader in the Department of Rural Development, Gandhigram Rural Institute, Deemed University and teaches Rural Development and Micro-Level Planning. Areas of specialisation are : (1) Cartography, (2) Geographical Information System, and (3) Remote Sensing. He has executed many projects on resource mapping, and watershed development funded by Central and State Governments. In particular, he prepared Resource Atlas of Dindigual District for National Informatics Centre, Planning Commission, Government of India. Presently he is engaged in evaluation of Integrated Waste Development Programme (IWDP) implemented by the District Rural Development Agency, Dindigual. He is a consultant to the IWDP programme of DRDA, of Dindigual and Karur districts, Tamil Nadu.

PERCEPTIONAL CARTOGRAPHY AND DISTRICT PLANNING

N.D. MANI
Reader, Department of Rural Development
Gandhigram Rural Institute (Deemed University)
Gandhigram, Tamil Nadu

CONCEPT PUBLISHING COMPANY, NEW DELHI-110059

ISBN 81-7022-892-1

First Published 2002

Published and Printed by
Ashok Kumar Mittal
Concept Publishing Company
A/15-16, Commercial Block, Mohan Garden
New Delhi-110059 (India)

Phones: 5648039, 5649024
Fax: 091-(11)-5648053
E-mail : publishing@conceptpub.com

ACKNOWLEDGEMENTS

I am deeply indebted to Dr. K.S. Gopalakrishnan, Professor and Head, Department of Geography, Madurai Kamaraj University and my guide for his able guidance, continuous encouragement and constant support throughout this study. I am grateful to him for all the pains he has taken in completing this research endeavour successfully.

I place on record my thanks to former Professor and Head, Department of Economics, Gandhigram Rural Institute (Deemed University) and convener of my Doctoral Committee and Dr. N. Sivagnanam, Professor and Head, Department of Geography, University of Madras and Member of my Doctoral Committee for their active role in shaping up this thesis in the present form.

I express my deep sense of gratitude to Mr. D.K. Oza, former Vice-Chancellor, Gandhigram Rural Institute (Deemed University) for his inspiration and encouragement for my research work. I am also grateful to Professor Haripad R. Subramonia Iyer, former Head, Department of Rural Development, Gandhigram Rural Institute (Deemed University) for his encouragement in pursuing my doctoral programme.

I owe my deepfelt gratitude to Dr. P. Arumukham, Professor and Head, Department of Rural Development, Gandhigram Rural Institute (Deemed University) for his whole hearted support and continued encouragement in this endeavour. I am thankful to Dr. A. Ramesh, former Professor and Head, Department of Geography, University of Madras for his constant support and advise in completing this work.

My sincere thanks to Dr. M.P. Boraian, Reader in Extension Education, Dr. P. Sumangala, Professor in Economics, Mr. C. Sivapragasam, Lecturer in Statistics, Dr. M.R. Kubendran, Reader in English, Dr. A. Joseph Durairaj, Reader in English, Dr. S. Ramasamy, Reader in Economics and Mrs. S. Lakshmi, Reader in English and Mr. B. Arunachalam, Assistant Registrar (Development) for their unstinted cooperation in improving the quality of the present work.

My sincere thanks are also due to Dr. K. Somasundaram, Professor and Head, Department of Computer Science, Gandhigram Rural Institute for his help and cooperation in using the computer facilities in cartographic work of the study. My special thanks are also due to the staff of the Computer Centre, Messrs. N. Palaniappan, I. Kaspar Raj, V. Petchi and S. Manokaran.

I am grateful to all the Heads of the Government Department in Dindigul District for helping me with necessary data and sharing their views and ideas for the present study. I wish to thank all the Heads of Geography Department in various Universities and Heads of Mapping Organization in India for their constructive criticism and valuable views on resource atlas.

My special thanks are due to Mr. C. Thirumavalavan, District Informatics Officer and Mr. S. Alagesan, District Informatics Assistant, National Informatics Centre, Dindigul District for extending all possible help in the data processing. I am also thankful to Messers. S. Manivel, P. Deivam, P.M. Chinnan and M. Lakshmanan for their help extended to me in this work.

Finally, I wish to record my deepfelt gratitude to all those who helped me in some way or other for the successful completion of this endeavour.

Gandhigram **N. D. MANI**

CONTENTS

LIST OF TABLES

LIST OF FIGURES

LIST OF MAPS

1

INTRODUCTION

1.1 INTRODUCTION

Planning has largely been a centralised exercise in India[1]. In order to reduce the disparities created by this approach and also to encourage participation of people in the planning process, efforts were initiated right from the First Five Year Plan to decentralise the planning endeavour. States like Maharashtra, Gujarat, Uttar Pradesh, and Karnataka have taken steps to operationalise decentralised planning at district level. Decentralised planning is mostly local-resource based and location/area specific. Hence, the success of the local planning depends upon the appraisal of local resources and their spatial distribution. For providing such information, a number of resource atlases are available at the national and state levels, but atlases at district level are hardly available. Consequently, the existing resource atlases are of less significance to the decentralised planners as they do not meet their requirements. The new challenges from the emerging vistases of technologies in the process of planning and new opportunities thrown up by the recent spurt in information technologies (like remote sensing) can be effectively bridged by cartography, which is an art, science and technology of map making. It is perceptional cartography that reconciles and harmonises the perceptions of map user and map maker. An attempt is made here to explain the recent development in Decentralised planning and Cartography and also their inter-dependence.

1.2 DECENTRALISED PLANNING

Decentralised planning has certain advantages over centralised planning. It enables a better perception of the needs of local areas, makes better informed decision making, gives people a greater voice in decisions concerning their development and welfare. It helps to achieve better coordination and integration among programmes. The felt needs of the people are effectively highlighted in decentralised planning. It ensures effective participation of the people, build up self-reliance by mobilising resources of the community in kind or money. It makes development self-sustaining and facilitates better exploitation of local resources. Growth potentials of the local areas can be identified and tapped more effectively in decentralised plan than in the centralised planning[2].

It is also considered as an effective means of reducing regional disparities and removing absolute poverty at the grass root level by encouraging people's participation[3]. As Kamta Prasad observes, the Planner at lower levels will have more intimate knowledge of the interdependence of activities and would be in a better position to develop integrated programmes to avoid duplication and can create maximum impact with minimum cost[4]. Moreover, it will reduce delays and ensure easy accountability as the programmes are implemented with people's participation and involvement. The rationale of decentralised planning is the size and diversity of the country, administrative convenience and efficiency. This type of planning not only ensures greater social justice but will also release latent potential of human and material resources. Thus, it should be regarded as not only a 'cost reducing' but also a 'resource augmenting process'[5].

1.2.1 Decentralised Planning Under Five Year Plans

The need for decentralised planning was stressed in all the Five Year Plans, in one way or the other. The First Five Year Plan emphasised the need for establishment of appropriate agency for development at the village level, deriving its authority from village community, integration of activities of various development departments in the district and linking up all self-governing institutions with the administering agencies of the state government[6]. The Second Five Year Plan discussed in greater detail the concept of local horizontal plan[7]. During the Second Five Year Plan foundation was laid for the establishment of three-tier system of Panchayati Raj Institutions (PRIs) for democratic decentralisation of planning process[8].

The Third Five Year Plan laid emphasis on 'PRIs' which would enable each area to realise its maximum development potential on the basis of local manpower and other resources, cooperative self-help and community effort, and effective use of the available resources and personnel[9]. A Study Team on Machinery for Planning was appointed to examine the district level planning in great detail. This team underlined the need for grass root planning for its ability to use the land and manpower in an effective manner, its capacity to evoke popular participation and its contribution to foster the democratic system[10].

During the Fourth Plan, the Planning Commission issued detailed guidelines for the formulation of district plans in 1969. It gave detailed account on specific steps to be followed in the formulation of the District Plan.

Since Fifth Five Year Plan, a number of special area programmes were implemented with the aim of alleviating poverty in rural areas. At this point of time, the need for decentralised planning efforts was increasingly recognised.

In 1977, a Committee on Panchayati Raj Institutions was appointed by the Planning Commission. It recommended the establishment of a strong *panchayat* system at the district level to respond both to the organisational needs within the district and to the demands of emerging linkages with the state[11].

The Sixth Plan formulated a central programme for strengthening the planning machinery at the district level.

A Working Group on District Planning (1982) was constituted to examine issues connected with decentralised planning. This working group reiterated the need for decentralised planning and also advocated a gradual approach towards introduction of district planning. It endorsed the view of bringing all planning activities at the district level under a single planning body, which would determine the priorities and allocate funds accordingly, among the various development sectors[12].

A Committee was appointed in 1985, to review the existing administrative arrangements for rural development and poverty alleviation programmes. It proposed the concept of district budget and stressed the need for strengthening the planning process at the district and block levels[13].

The Seventh Five Year Plan, while endorsing the recommendations of the working group on district planning, urged the states to take steps in four important aspects of a decentralised district level planning *viz.*, functional decentralisation, financial decentralisation, appropriate planning mechanism at the district level and appropriate budgeting and re-appropriation procedure[14].

In 1990, Planning Commission issued the guidelines on District

Planning prepared by the National Institute of Rural Development (NIRD), Hyderabad. Details of the procedures to be followed in the planning of individual sectors like agriculture, horticulture, forestry, animal husbandry, industry, transportation, electricity, water supply, health, education, housing, general settlement plan including budgeting and financial planning were explained in the guidelines. It opined that district budget ought to be an agglomeration of sectoral budgets[15]. It is the latest effort taken by the Planning Commission to operationalise district planning.

The reports of the committees, working groups and study teams had a considerable impact on the decentralisation of planning in India. Many states have taken steps for decentralising the planning process at the lower levels particularly at district level.

1.2.2 District Planning in India

In India, sub-state level planning is largely referred as district level planning, necessarily because a large volume of data on various themes for a long period of time and administrative infrastructure in terms of functionaries of various development departments are available at district level. Also, district is considered large enough to be a viable economy for planning[16]. Almost every state has decentralised some planning function and created some organisation for planning at district level. However, only a few states like Maharashtra, Gujarat, Uttar Pradesh and Karnataka have allotted staff, allocated funds and delegated adequate powers to make district administration an effective planning agency. The major functions of these district level planning bodies are coordination, monitoring, review, finalisation of Annual and Five Year Plans and data feeding to the higher levels.

1.3 FACETS OF DISTRICT PLANNING

Planning at the District level has three significant facets, viz., 1. Social, 2. Economic, and 3. Spatial.

Social facet refers to the society, for whose development the plan is drafted, economic facet denotes the economics of the plan and the spatial facet explains the planning area or region in which the target people live and where resources are available. The first two facets become meaningful only when localised and seen in their spatial perspectives as they are distributed over an area.

To take care of spatial facet of district planning, the Working Group on District Planning, included spatial planning as one of the three components of the core of District Planning, the other two being project/scheme planning, manpower and employment planning[17].

1.3.1 Spatial Planning

Space is a common denominator in all human activities, and therefore, spatial planning cuts across and integrates all human activities[18]. It is easy to operate a sectoral plan within a spatial framework but it is difficult to do the opposite. Moreover, human activities and natural resources are often maldistributed. "To minimise the hardship of maldistribution, we have to plan for a new structure of human activities offsetting the natural disadvantage of the backward areas"[19]. It is, therefore, now acknowledged that there exist a close interdependence between national development and spatial pattern of human activities[20]. It is necessary to plan to reorganise the spatial structure of these activities in order to achieve the basic social, economic and environmental goals of planning. Without the change in the spatial structure, a forward move in one sector may retard progress in another. The main objective of spatial planning is to formulate a location-allocation strategy for guiding new investment on various services and activities at the appropriate location to ensure equitable and efficient spatial distribution[21]. In other words, spatial planning implies focussing of investment of socio-economic infrastructure at certain places which are selected in terms of maximising access to the beneficiaries at the minimum cost[22]. It is, therefore, locational planning for social facilities and developmental infrastructure, needed keeping in view the actual requirements, problems and needs of people living in an area[23].

There are two types of spatial planning viz., adoptive and active[24]. The adoptive spatial planning is more or less passive planning and it refers to adjustment with existing locational pattern. In this type an attempt is made to streamline the process of spatial evolution so as to achieve a spatial system which matches the requirements of a production function for efficiency of growth[25]. The active spatial planning starts with the recognition of interplay and feed back between economic development and spatial process and it, therefore, seeks to achieve within a dynamic and historical context a pattern of evolution of the spatial structures judged to be most efficient for promoting a sustained process of economic development[26].

1.3.2 Significance of Maps in Spatial Planning

For spatial planning, information on all aspects of the resource endowment, resource utilisation levels, growth trends and social structure with regard to area under planning is required[27]. Very often, data which are available are not location specific. Plans formulated on the basis of such inadequate data result in faulty prioritisation and allocation.

One of the pre-conditions for the integrated development of an area in multi level planning system is spatially related resource information. Generation and documentation of resource data at all levels of planning is an imperative. In this context, the role of maps cannot be over emphasized because the availability of resources in an area can be identified and assessed scientifically by using the maps, which are in modern times, supplemented by aerial photos and satellite imageries. Maps can be used to document resource data for deriving resource information at a later date. They act as an effective media to transmit information to the planners. Maps, as a tool in planning, have many functions to do as listed below:

A map has advantage over text. No verbal description can reveal the impact and retention possibilities of a map. It gives a synoptic as well as panoptic view of the planning area.

Resource data can be represented in map giving proper spatial treatment. Thus, the utility value of the original data will be enhanced.

Maps also act as a laboratory by way of bringing in a miniature model of planning area in two as well as three dimensions on one's table and this minimises visits to the actual area of planning in its initial stages.

Apart from physical features, maps can also be designed to portray socio-economic factors bringing out the relationships between and among them. When a set of maps are drawn at regular intervals on various resources of an area, they serve as a source of resource data/ information for planning. The changes in resource utilisation pattern over space and time can be brought out by these maps.

Maps drawn with the available data can be put to a variety of uses. They can be used to forecast change in future, measure the anticipated changes and provide alternative cost-effective models for choosing the most appropriate one. For identification and delimitation of developed, developing and under-developed areas, problem areas, clusters, thrust areas etc., in a planning region, the only usable and the most effective tool is a thematic map.

1.4 RELEVANCE OF CARTOGRAPHY IN PLANNING

Cartography — the art, science and technology of map-making has played a pivotal role in the formulation of development plans in the country by preparing maps to the planners. As a result of data explosion and experience gained, as part of the planned development a wide variety of data are found in many Government Departments at different levels. This wealth of data has been further augmented by recent developments in remote sensing. These data are so massive that they could not be fully used in conventional tabular form. Therefore, use of cartographic methods can provide better visual presentation of the data. It is because of this importance, cartography has been developed in independent India at rapid pace through creation of many specialised agencies and these institutions by their constant endeavours, have made India one of the best mapped countries in the world[28].

1.4.1 Major Organisations Involved in the Preparation of Maps/ Atlases

The Survey of India is the premier mapping organisation in India. It is a national organisation for survey, mapping and research and development in cartography. Geological Survey of India is another oldest scientific institute, which is responsible for geological and mineral survey and publication of their maps. National Atlas Organisation, at the time of its inception in 1956, was exclusively preparing thematic maps and atlases at the national level. Indian Marine Survey Department was established at Calcutta, now renamed as Naval Hydrographic Office and shifted to Dehra Dun, carries out hydrographic surveys of the Indian coasts and harbours and publishes nautical charts and related documents, required for navigation and other purposes. An independent agency *viz.*, National Remote Sensing Agency, established during 1975 in Hyderabad, prepares thematic maps based on aerial photographs and satellite imageries. Registrar General of Census Operations also produces population maps and atlases of India by using census data.

In addition to the above institutions Indian Meteorological Department, National Geophysical Research Institute, Forest Survey of India, Office of the Directorate of Military Survey and the Tourism Departments of all states publish atlases/maps. The Survey and Land Records Department in each state produces cadastral maps for their respective states. Recently, geography departments of some Indian universities like Osmania and Madras have started publishing state level atlases.

1.4.2 Perceptional Cartography

Perception in general is a process whereby sensory stimulation is translated into organised experience[29]. It is also explained as recognition or identification of something[30].

A branch of cartography that deals with perceptional aspect is known as Perceptional Cartography. It is concerned with, analysis of salient features in the reality and creation of the graphic document which would allow another person (with or without previous knowledge of the reality) to reconstruct these features sufficiently well as to allow him to establish a relationship so as to enable him to locate himself precisely in that reality[31]. Therefore, perceptional cartography is concerned with the mental reaction of the map user and cartographer to the reality and its representation through maps.

Cartographic products should meet two conditions, if they are to serve the planners. Firstly, contents of the cartographic work should be relevant to the needs of the users. Secondly, the method of presentation should be understandable to him. Thus, the needs and capabilities of the users ought to determine the work of the cartographer. Therefore, even maps and charts carrying relevant information, can serve a planner only when he is able to visualise and understand what has been visualised and presented by the cartographer. However, many of the cartographer/mapping organisations do not seem to give importance to the needs, interests and abilities of the user. Therefore, Arunachalam observes that cartography has two clear-cut and distinct perspectives, one from a map maker's point of view and another from a user's point of view. Unless the two perspectives are well blended and integrated, a cohesive perspective of the map as a whole cannot be effected[32]. In order to achieve the maximum appreciation by the map users, cartographer has to carefully investigate the purposes served by the map and environmental conditions in which the maps are used[33]. In other words, the creative work of the cartographer should be based on the needs, interests and subjective condition of the map user. Hence, the need for perceptional cartography.

The above analysis clearly depicts the role of cartography in the planning process and portray their inter-dependences. The study aims at extending the concept of perceptional cartography to the maps required for district level planning (Fig. 1.1). A review of literature will

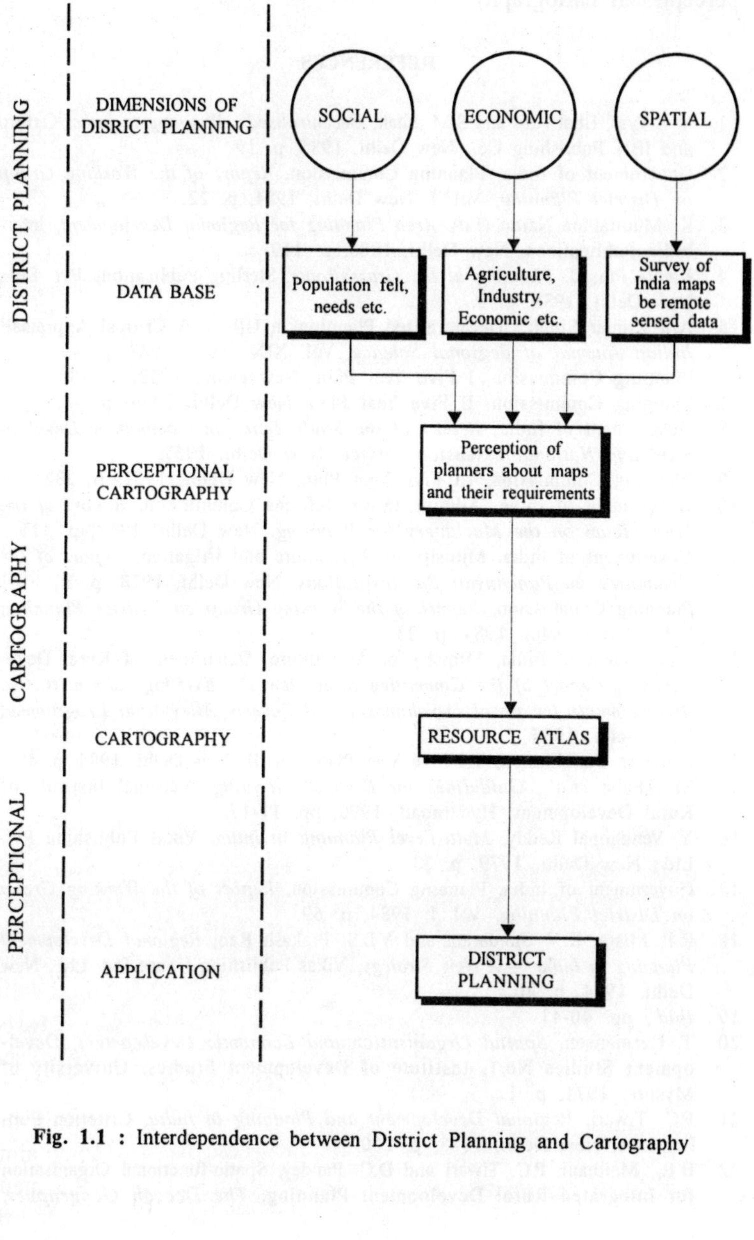

Fig. 1.1 : Interdependence between District Planning and Cartography

enable us to ascertain the existing level of development in the subject and identify gaps, if any, therein facilitating deeper investigation in perceptional cartography.

REFERENCES

1. V. S.Vyas, Bhat Anil and S.M. Shah, *Decentralised Planning in India,* Oxford and IBH Publishing Co., New Delhi, 1985, p. 19.
2. Government of India, Planning Commission, *Report of the Working Group on District Planning,* Vol. I, New Delhi, 1984, p. 22.
3. K. Munirathna Naidu (Ed). *Area Planning for Regional Development,* Inter-India Publications, New Delhi, 1983, p. 119.
4. Kamta Prasad, *Planning at the Grass Roots*, Sterling Publications Pvt. Ltd., New Delhi, 1988, p. 9.
5. Ajit Kumar Singh, 'Decentralised Planning in UP — A Critical Appraisal' *Indian Journal of Regional Science*, Vol. XIX, No.1, 1987, p. 49.
6. Planning Commission, I Five Year Plan, New Delhi, 1952, p. 48.
7. Planning Commission, II Five Year Plan, New Delhi, 1956, p. 48.
8. Government of India, *Report of the Study Team on Community Development and National Extension Service*, New Delhi, 1957.
9. Planning Commission, III Five Year Plan, New Delhi, 1961, p. 332.
10. Government of India, Administrative Reforms Commission, *Report of the Study Team on the Machinery for Planning,* New Delhi, 1967, p. 115.
11. Government of India, Ministry of Agriculture and Irrigation, *Report of the Committee on Panchayati Raj Institutions,* New Delhi, 1978, p. 8.
12. Planning Commission, *Report of the Working Group on District Planning,* Vol. I, New Delhi, 1984, p. 23.
13. Government of India, Ministry of Agriculture, Department of Rural Development, *Report of the Committee to Review the Existing Administrative Arrangements for Rural Development and Poverty Alleviation Programme,* New Delhi, 1985.
14. Planning Commission, VII Five Year Plan, Vol. II, New Delhi, 1985, p. 403.
15. M. Thaha *et.al., Guidelines for District Planning,* National Institute of Rural Development, Hyderabad, 1990, pp. 11-17.
16. Y. Venugopal Reddy, *Multi-Level Planning in India,* Vikas Publishing Pvt. Ltd., New Delhi, 1979, p. 83.
17. Government of India, Planning Commission, *Report of the Working Group on District Planning,* Vol. I, 1984, p. 69.
18. R.P. Misra, K.V. Sundaram and V.L.S. Prakash Rao, *Regional Development Planning in India — A New Strategy*, Vikas Publishing House Pvt. Ltd., New Delhi, 1974, p. 40.
19. *Ibid.,* pp. 40-41.
20. T. Hermansen, *Spatial Organisation and Economic Development,* Development Studies No.1, Institute of Development Studies, University of Mysore, 1971, p. 1.
21. P.C. Tiwari, *Regional Development and Planning in India*, Criterion Publications, New Delhi, 1988, p. 200.
22. B.P. Maithani, P.C. Tiwari and D.C. Pandey, Spatio-functional Organisation for Integrated Rural Development Planning, *The Deccan Geographer,*

Secundarabad, 1982, pp. 145-163.

23. P.C. Tiwari, *Op.cit.*, p. 13.
24. T. Harmensen, *Op.cit.*, p. 15.
25. P.C. Tiwari, *Op.cit.*, p. 13.
26. B.P. Maithani, *Spatial Analysis in Micro Level Planning*, Omsons Publications, Guwahati, 1986, p. 231.
27. Government of India, Planning Commission, *Report of the Working Group on District Planning*, Vol. I, New Delhi, 1984, p. 72.
28. S.P. Das Gupta, Growth and Development of Cartography in India, *Indian Cartographer,* Vol. IV, Hyderabad, 1982, p. 19.
29. David H. Bailey *et al.*, (Eds). *Encyclopaedia Britanica,* McGraw Hill Book Co., New York, 1979, p. 386.
30. Human Friedhoff, *et al.*, (Eds). *The University Desk Encyclopaedia,* E.P. Dulton Inc, New York, 1977, p. 783.
31. John C. Bartholomew and Ian A.C.Kinniburgh, The Factors of Awareness, *Cartographic Journal,* Vol. 10, No. 1, 1973, p. 60.
32. B. Arunachalam, Status of Cartography Education in India and Measures for its Quantitative Improvement, *Indian Cartographer,* Vol. VII, 1987, p. 333.
33. A.Kolancny, Cartographic Information — A Fundamental Concept and Term in Modern Cartography, *Cartographic Journal,* Vol. 6, No. 1, 1969, pp. 47-49.

2

REVIEW OF LITERATURE AND METHODOLOGY

2.1 INTRODUCTION

This chapter is devoted to review of literature and designing of methodology. Perusal of literature consisting of atlases/maps (produced in India) and research work in perceptional cartography has enabled the researcher to examine the potentials and identify problems in the use of resource atlases/maps for planning purposes. Thus, the review of literature facilitated the identification of problems and formulation of objectives for the present study.

The methodology covers identification of respondents, selection of sample respondents, sources and instruments for collection of data and statistical tools of analysis. Chapterisation of the study forms concluding part of this chapter.

2.2 REVIEW OF LITERATURE

2.2.1 Review of Literature in Atlases/Maps

As already noted, India is one of the best mapped countries in the world. It was the establishment of National Atlas Organisation (NAO) that gave great impetus to the production of thematic atlas in India. Therefore, the review of atlases starts with perusal of National Atlas of India brought out by NAO in Hindi comprising 26 thematic maps on India[1].

Indian Agricultural Atlas was published with 53 generalised maps focusing on distribution of cropped land, area under different crops, world distribution of cropped land, area under rice, wheat, sugarcane, jute and cotton, livestock, agricultural implements, forests, physical features, climate, soil, administrative divisions and transport[2]. The Director of Economics and Statistics, Bhopal, brought out Economic and Social Atlas of Madhya Pradesh, comprising mainly graphs and charts[3].

The Government of Mysore, published Mysore State in Maps. It contained maps of the state with pictorial diagrams representing different sets of socio-economic data[4]. Indian Oilseeds Atlas was published by Indian Central Oilseeds Committee to show the production of groundnut, rape, sesame, linseed, and castor and major oilseeds growing states[5]. Indian Lac Cess Committee brought out an Atlas of Indian Lac, containing 16 maps, of which two are World maps showing distribution of lacciferidae and its hypothetical range and major shipping routes, one map of Asia depicting lac growing countries, four maps of India showing distribution of spices and lachosts, areas of lac production and state broodfarms, and two maps of each of West Bengal, Bihar, Madhya Pradesh, Orissa, Uttar Pradesh, Maharashtra, one showing area of lac production and the other depicting important lac centres[6].

A number of single theme atlases were brought out in 1959. Coconut Atlas of India, contained 26 maps, mostly of pictoral type[7]. Indian Tobacco Atlas comprised 18 maps of India, representing physical features, tobacco production, tobacco marketing centres, trading centres, research centres and centres of exploratory schemes[8]. Indian Cotton Atlas was published to show the different aspects of cotton. It has eight sections showing physical features, soil, communication, area under irrigated cotton and area under long, medium and short staple cotton and the progress of improved variety of cotton, centres of cotton research, cotton seed distribution and protected areas for cotton and state maps showing different varieties of cotton grown[9]. Indian Jute Atlas comprised three World maps showing jute growing centres, distribution of jute mills, export of jute from India and two maps of India, of which one showing generalised transport and the other representing distribution of jute areas confined to the eastern zone comprising mainly Assam, Bihar, West Bengal and Orissa, five maps of the jute growing States, research station and seed farm, rainfall in jute growing areas, soils of jutė region and inland transport[10]. Agricultural Atlas of Madhya Pradesh was published in 1959 containing mainly graphs and charts[11]. Statistical Atlas of Orissa was brought out to show rainfall, population density, area and yield of rice, sugarcane etc., and community development[12]. A number of economic maps were included in the Statistical Atlas of Rajasthan[13]. During the same period, Bureau of Economics and Statistics of Government of Andhra Pradesh published Statistical Atlas of Andhra State[14]. National Institute of Nutrition of Indian Council of Medical Research published Nutrition Atlas of India[15], and Diet Atlas of India[16]. The aim was to document the valuable data scattered in several publications. Atlas of Northern Frontier of India was brought out by the Government of India, comprising reproduction of 39 maps published on India, China, Britain, France and Germany. They are arranged in four sections *viz.*, General, Ladakh section, Assam section and Chinese maps of 1928[17].

The first atlas with all requirements of the maps for planning was brought out *viz.*, Mysore state: An Atlas of Resources by L.T.A. Learmonth and L.S. Bhat. It contained 120 maps in black and white, with a note on sources and techniques[18]. In 1962, the National Council for Applied Economics Research, brought out Economic Atlas of Madras State[19]. In 1966, an Agricultural Atlas of Punjab was published[20]. The Joint Committee for diagnostic survey of Damodhar Valley, set up by the Ministry of Power and Irrigation brought out Planning Atlas of Damodhar Valley Region. It contained 32 plates depicting physical, socio-economic parameters of the valley[21]. Calcutta Metropolitan Planning Organisation, prepared Burdwan District Atlas[22].

The Agricultural Atlas of Rajasthan was brought out by S.K. Sen[23] and the Registrar General of Census of India published a Census Atlas of India based on the 1961 data[24]. Indian Meteorological Department brought out an abridged edition of Climatological Atlas of India[25]. Besides, a separate Rainfall Atlas of Indian consisting of 61 maps[26]. Based on the census data of 1971, a series of Administrative Atlases of Haryana[27], Tamil Nadu[28], Rajasthan[29] and Andhra Pradesh[30] were brought out by the Registrar General of Census of India.

National Atlas and Thematic Mapping Organisation (NATMO) formerly known as National Atlas Organisation (NAO) published the Irrigation Atlas of India to cater to the needs of irrigation planners. It formed a part of the report submitted by the Irrigation Commission to the Government of India[31]. P.K.Tiwari of G.B.Pant University of Agriculture and Technology brought out Agricultural Atlas of Uttar Pradesh to highlight the agricultural resources of the State[32]. At this point of time, B.P. Amasthya published a combined reprint volume of James Rennell's Memoir of a Map of Hindustan and the Bengal Atlas[33]. Osmania University published Planning Atlas of Andhra Pradesh in collaboration with the Survey of India[34]. The production of maps of a State for planning purposes paved the way for production of State level resource atlas. J. Singh brought out an Agricultural Atlas of India.[35] NATMO published Tourist Atlas of India comprising 25 plates for the use of foreign and local tourists[36]. Department of Agriculture, Government of India brought out Indian Agricultural Atlas[37]. A Statistical Atlas of the Union Territory of Pondicherry comprising Pondicherry,

Karaikal, Mahe and Yanam regions was prepared by the Bureau of Statistics and Valuation, Government of Pondicherry. The atlas consists of maps on network of rivers, land, population, land use, agricultural production, animal husbandry, fisheries, medical and public health etc.[38]

NATMO brought out an Atlas of Forest Resources of India consisting of 36 plates designed to portray the present status of forestry in India, with a view to help conservation and development planning of forests[39]. An Atlas on Kerala in Maps was brought out, covering all sectors of economy with explanations and statistical tables[40]. Indian Meteorological Department brought out Agro-climatic Atlas of India[41]. The Census of India, to mark the centenary of Indian Census: 1882-1971, published Indian Census Centenary Atlas[42]. Indian Literacy Atlas was published by Central Institute of Indian Languages[43]. NATMO brought out an Atlas of Agricultural Resource of India comprising 36 multi-coloured maps.[44]

The English edition of National Atlas of India in eight volumes was brought out by NATMO, comprising 300 plates. The eight volumes dealt with General and Political, Physical and Geomorphology, Climate and Bio-Geographical Factors, Population and Transport, Landuse and Industrial Activities, Agriculture and Economic Activities, Social and Regional Maps and Historical and Cultural details.[45]

N.B.K. Reddy of Sri Venkateswara University prepared Rayalaseema Resource Atlas, as part of his project though it was not published. In the year 1983, A. Ramesh and P.S.Tiwari of University of Madras, published Basic Resource Atlas of Tamilnadu to highlight the nature of resources and the pattern of resource use in Tamil Nadu. The atlas contained maps on land resources, water resources, biotic resources, mineral resources, human resource, irrigational aspects, agricultural development, industrial development, development infrastructure and regional resource development[46].

The Resource Atlas of Kerala brought out by Centre for Earth Science Studies, Trivandrum, presented the basic resource information available in different State and Central agencies through 132 multi-coloured maps in 58 plates. A wide spectrum of themes *viz.,* natural resources, human resources, resource development and also the status of resources and their utilization were depicted.[47]

The Social and Economic Atlas of India was published by Oxford University Press. It provided an up-to-date picture of India's habitat, society and economy through maps, charts, tables supplemented by a brief text. It also looks at India's development potential in the light of the Seventh Five Year Plan and examined what the economic and social scene is likely to be in India in nineties. All these pieces of information were presented in nine sections *viz.*, land, people, climate, natural resources, international equations, infrastructure, produce, tourist vista and national economy[48].

Planning Atlas of Medak District was published by the Centre for Economic and Social Studies, Hyderabad[49]. This atlas contains 160 maps presented in 57 plates. These plates have been arranged in eight sections. The first section shows location, while the second section deals with physical setting. The third and fourth sections show details of agriculture and irrigation respectively. While section five delineates pollution, section six depicts socio-economic infrastructure. Plates on rural development programmes and industrial development are presented in sections seventh and eighth respectively. Survey of India has taken up a pilot project on 'District Maps' with an aim of applying cartography in collection, preparation and presentation of data for district planning upto block level. In this project, it is proposed to collect data from various Central and State organisations and represent in a set of insets[50].

Besides these atlases, various planning agencies prepared maps for their respective areas of specific requirements. L.S. Bhat *et al.*, conducted a case study on Central place hierarchy in a developing agricultural region. In this study, regional structure at district and *tahsil* levels were studied with appropriate scales of map (one inch to four miles and one inch to one mile scale). The map on physical landscape and its spatial variation provided the clue to the variations in land use and productivity and also in emerging pattern of transport and growth of settlement as nodal centre. Data on land use, cropping pattern and demography were mapped for further planning. The outcome of this study was used to develop strategies in district planning in terms of its area and locational components[51].

An attempt has been made by P.C. Tiwari, to examine the applicability of theoretical aspects of spatial organisation by applying the spatial planning approach at District level[52]. In this study, maps of Survey of India were used in order to generate data and prepare maps on relief, water bodies, forest, agriculture, settlements, transport linkages etc. Maps have also been drawn to study the pattern of spatial interaction. In addition, maps prepared on various aspects by government departments were also used. The findings were used to design a spatial framework in such a way that the spatial and functional gaps would be bridged and the regional imbalances in the level of socio-economic development removed.

In 1969, a Planning and Development Board was established for Rayalaseema area of Andhra Pradesh. One of the methodologies followed in preparing the plan for the area was mapping of activities and zoning. To tackle spatial elements and also to correlate soil, climate, irrigation and cropping pattern with each other particularly settlement hierarchy maps were superimposed and analysed[53]. In Gujarat, maps and other cartographic techniques are employed for analysing the common issues and for better development planning at district level[54].

The above review of atlases and maps shows that in the beginning, the atlases were prepared at the national level covering all resources, subsequently they were followed by mono theme resource atlases and state level resource atlases. For a country of India's size and diversity, a single district level atlas – Planning Atlas of Medak District – is a meagre source of guidance which needs further enrichment. Hence, it is imperative to evolve a design for preparing resource atlas at the district level which can be readily used for planning purposes and general understanding of the resources of the country. Precisely to meet this requirement, the present investigation seeks to prepare a comprehensive list of resource maps at district level for planning purposes.

Several studies have been conducted in perceptional cartography. They are reviewed here for the purpose of understanding the present status of perceptional cartography.

2.2.2 Review of Literature in Perceptional Cartography

The review of research studies in the field of perceptional cartography is given below :

De. Brommer[55], Kamkov[56], Nikishov and Preobrazhensky[57], Ratjaski[58] have studied the problems of improving the effectiveness of map communication through standardization of map symbols. Mackay[59], Clarke[60], Ekman and Junge[61], Flannery[62], Meinhoefer[63,64], Crawford[65] have studied the accuracy, error, variability of the values of symbols like bars, circles, squares, spheres etc.

Phillips and Noyes have assessed the effect of removing topographic symbols which are only of minor importance to the map reader.[66] Mark and Edward have conducted a study to derive and evaluate a topographic map symbology that would support the performance of basic map using tasks[67].

The use of colour, visibility etc., was studies by a group of cartographers. Taylor considered the visibility of different coloured printed text against background of other colours[68]. Cuff has studied the use of colour for showing 'temperature regions' and reaction of students to it for designing maps for atlases and text books[69]. Patton and Crawford have studied the use of standard sequence of colour in hypsometric maps and transmission of information to the user[70].

Bartz has investigated the aspect of legibility in cartographic typography[71]. Gerber found out which of the selected typefaces would be the easiest to read and which would be preferred by the school children[72].

Taylor has identified the differences between the map perceived by users and tried to deduce design requirements. The subjects were 26 pilots and 10 navigators and the study involved 18 topo maps at 1:250000 and 1:253440 scales. The findings were that none of the maps was rated ideal by the subjects. Pilots preferred maps with bold, distinctive layer tints, colour for relief and both the groups required maps with reduced information content and less visual clutter[73].

Phillips *et al.,* has tested the legibility of four different types of relief maps. The maps were compared with the help of 13 different map reading test questions posed to 179 subjects. The result was no single type of map was the best for all purposes[74]. Chang used a series of five choropleth maps using the same data but different classings to study visual aspects of class intervals in choropleth mapping. The subjects were to rank the maps in the order of preference for the final map production and provide explanation for their ranking[75].

William has tried to establish 'equal appearing intervals' for printed shading screens. By using subjects' opinion of what appeared to be equal steps in change of darkness, he constructed a curve of the grey spectrum as a guide to the choice of shading screens to be used on maps[76]. Jenks and Knos[77] and Strossel[78] conducted studies similar to Williams. Cuff observed that none of the preceeding attempts to set up curves of the grey spectrum had been concerned with whether or not map readers associated map tones with higher or greater amounts, but had merely sought equal appearing intervals[79]. Muller used the criteria to compare the looks and patterns of choropleth map[80]. Monmonier studied the selection of class intervals on choropleth maps with an aim of maximising the visual correspondence between the variables to be categorised and a reference variable with which it associated[81]. Jenks and Caspall enquired into how much vertical exaggeration would be needed on three dimensional diagrams to make them look realistic. The subjects were experienced geographers, geologists and engineers, who use topographical maps in their work. It was found that exaggeration is required and that desired ratios of exaggeration varied with the degree of relative relief[82]. De Lucia studied the effects of shaded relief in comparison with contour lines with regard to accessibility of other map information[83]. Head collected the responses of map readers to a variety of techniques for land-water differentiation[84].

Phillips, De Lucia and Skelton attempted to compare the information carrying capacity of several different types of area land form symbols[85]. Jenks and Caspall studied the relationship between map accuracy and information carrying capacity of choropleth map[86]. Brandes conducted a study similar to that of Phillips but compared several additional symbolisation and inquired of different types of information to be communicated[87].

Alan and Donald tried to re-design standard legend formats so that legend symbols would be seen in a visual context more closely resembling their actual situation within the body of the map and significantly improve the efficiency of map information processing[88].

There is another approach in the field of perceptional cartography using the perception of users about their preferences, needs etc., for improving and re-designing the map. Sheppard and Adams conducted a questionnaire survey which included 128 drivers for map reading tasks, using two maps published by two different agencies. The survey was about which of the two maps was preferred and about the features required in a road map[89]. One of the few other studies to include user information in recreational map planning is that by McGrath and Castner. Their survey of users however, was undertaken prior to the design of the recreational map. The study included a preliminary survey of users within a park and subsequently a recreation map was designed and published[90]. A study similar in concept was conducted by Patrick. According to him, creation of maps for children requires that the map designer should be aware of the abilities, needs and attitudes of the child. The content and structure of the map were decided by using 300 young children as subjects[91].

In India, researchers have not evinced interest in studying symbols or individual elements of the map. However, a trend in this direction was set in early 1980's. The Indian National Cartographic Association constituted a Commission for School Atlas to finalize the various aspects of two separate atlases for students of class II to VIII and one senior level atlas to serve the purpose of the students of high school level. Accordingly, a questionnaire was sent to various agencies to find out the requirements to be met by these atlases. With this information and also on the basis of various atlases available, the list of maps/themes to be included in the atlases were prepared. The report of the commission also gives the details like size, scale, projection, symbols and layout of maps[92]. S.K. Sahni conducted a study of re-design and construct topographic map on scale 1:25000 to make it more multipurpose. The data on different maps used by the user in their departmental activities, information generated from the maps, and also additional information required were collected. Based on this, a sample map was produced with modern technique of scribing. As the users gave a long list of details for incorporation, only two or three demands of each were incorporated in the sample map. By keeping this map as a reference map, the author suggested a set of overlays and footnotes for adding the data on map proper. There were 30 new additions in the map, with modification and get-up of the map improved in the succeeding editions[93].

In foreign countries, significant contribution was made on various aspects of symbols, use of colour, maps carrying capacity of information etc. Thus most of the studies tended to be highly specialised and wholly directed at investigating and testing the user's reaction to one particular element of the map. A small group of researchers, designed the entire map based on the perception of the users.

In India, only a few studies are reported on perceptional research on various components of the map like legend, title, symbols etc. However, in one of the studies only one map has been designed based on the requirements of the users.

The above review of literature on atlases and research in perceptional cartography highlights the need for drawing resource maps at district level for planning purposes on the basis of the evaluation of available resource atlases and requirement of the user of district level planning.

2.3 METHODOLOGY

2.3.1 Statement of the Problem

The importance of district level planning is bound to increase in the days to come and will be feasible only with the availability of appropriate resource atlas which will be indispensable for fulfilment of the spatial imperatives of decentralised planning.

One of the pre-requisites of district level planning is the availability of data for decision making and the existing data is seldom available at the right time and place. Also, whatever meagre data available are mostly outdated and inadequate. Hence, there is need for a systematic and scientific collection, documentation and representation of data for the present and future use at the district level for planning purposes.

The data as required by the planner would be both spatial and statistical. Most of the time, the planners have to integrate these two sets of data for deriving useful information and for taking right decisions. The mapping organisations publish atlases covering all sectors of economy at the national and state levels. However, there are only a few district level resource atlases available with the Planning Atlas of Medak District being an exception. Thus, there is a wide gap between the available resource atlases and a resource atlas as required by district level planners. The gap is in the content and also in the level of resource information. The resource atlases which are now available at national and state levels will have to be supplemented by resource atlases at district level.

For preparation of appropriate resource atlases at the micro level, the perception of planners at district level about various aspects of use of existing maps and details of a resource atlas are considered essential. Technical feasibility of adopting the perception of the map users may have to be ascertained from the map designers. Hence, in the present investigation, an attempt has been made to ascertain the perception and requirement of district level planners and map designers on their requirements and norms of a model resource atlas for planning purposes.

2.3.2 Title of the Study

The present study is titled as "An Exercise in Perceptional Cartography –Evaluation of Resource Atlases and their Application to District Planning".

2.3.3 Study Area

The study covers Dindigul District, Tamil Nadu for the purpose of data collection and analysis of perceptions and views of map users and for drawing model maps. This district has been selected for the study as it happened to be the place of work of the researcher.

This district is one of the 29 districts of Tamil Nadu State. It has 304 village panchayats and 24 town panchayats. Revenue village numbering 347 are grouped under 6 *taluks*. There are 14 development blocks and two municipalities present in the district. Total population of 17,60,601 is distributed over 6,480.18 sq. km of geographical area.

2.3.4 Objectives of the Study

The prime objectives of the study are :

1. to identify the sources of data for district planning and problems therein,
2. to study the existing information system at the district level,
3. to evaluate selected resource atlases and guidelines on district planning issued by the Planning Commission, Government of India,
4. to portray the perception of map users and views of map designers,
5. to evolve a list of map themes for district level planning on the basis of evaluation of resource atlases and the perception of map users and designers, and
6. to prepare a set of model resource maps on selected themes for Dindigul District, Tamil Nadu.

2.3.5 Operationalisation of Terms

Map User: Officers heading various departments at district level, who are either actual or potential map users.

Map Designer: Heads of Department of Geography in Indian Universities and the Heads of Mapping Organisation in India.

2.3.6 Methodology

The proposed exercise has four components *viz.*,

(a) evaluation of resource atlases,

(b) feasibility of adopting them as a model for district level planning,

(c) preparation of list of map themes for a resource atlas for district level planning purposes on the basis of evaluation and adoption as proposed, and

d) preparation of a model resource maps on selected themes for Dindigul District, Tamil Nadu.

For the purpose of evaluation of resource atlases, National Atlas of India at the national level and the Basic Resource Atlas of Tamil Nadu and

Resource Atlas of Kerala at the state level also the Planning Atlas of Medak District at the district level were chosen. Norms for the evaluation of the resource atlas were, relevance of themes to district level planning, level of data (village-wise, block-wise etc.) represented in the map, scale and technique of the mapping (chorochromatic, choropleth etc.).

In the review, themes of many of the maps were found to be potentially useful for district level planning even though such maps require certain modifications and additional themes. This set the researcher in search of new themes for maps as needed in district planning. Other sources of map themes for resource atlas were guidelines for district planning issued by the Planning Commission, Government of India in 1969, 1984 and 1990. Examination of above atlases and guidelines enabled the researcher to prepare 258 themes of map for district level planning.

These themes of map were circulated among 40 district level officers and 63 map designers for their comments on the relevance of map themes, scale of map, level of data and periodicity of revision. On the basis of data collected from two categories of respondents, an endeavour has been made to prepare a consolidated list of themes of resource maps as relevant for district level planning and draw a set of resource maps for Dindigul District (Fig. 2.1).

2.3.7 Respondents

As already mentioned under methodology, the study involved collection of view from two sets of respondents viz., map users, heads of Government departments functioning in a district and map designers, comprising professors of geography in Indian Universities and heads of mapping organisations.

2.3.8 Selection of Respondents

The following list of 40 Government offices functioning at the district level was selected for the study in consultation with the officials of collectorate of Dindigul District, Dindigul.

1. Agri. Engineering (Soil Conservation Scheme)
2. Agri. Engineering (Western Ghats Development)
3. Public Works
4. Highways and Rural Works
5. Ground Water
6. Tamil Nadu Water and Drainage Board (Rural Water Supply)
7. Assistant Director of Statistics
8. National Informatics Centre

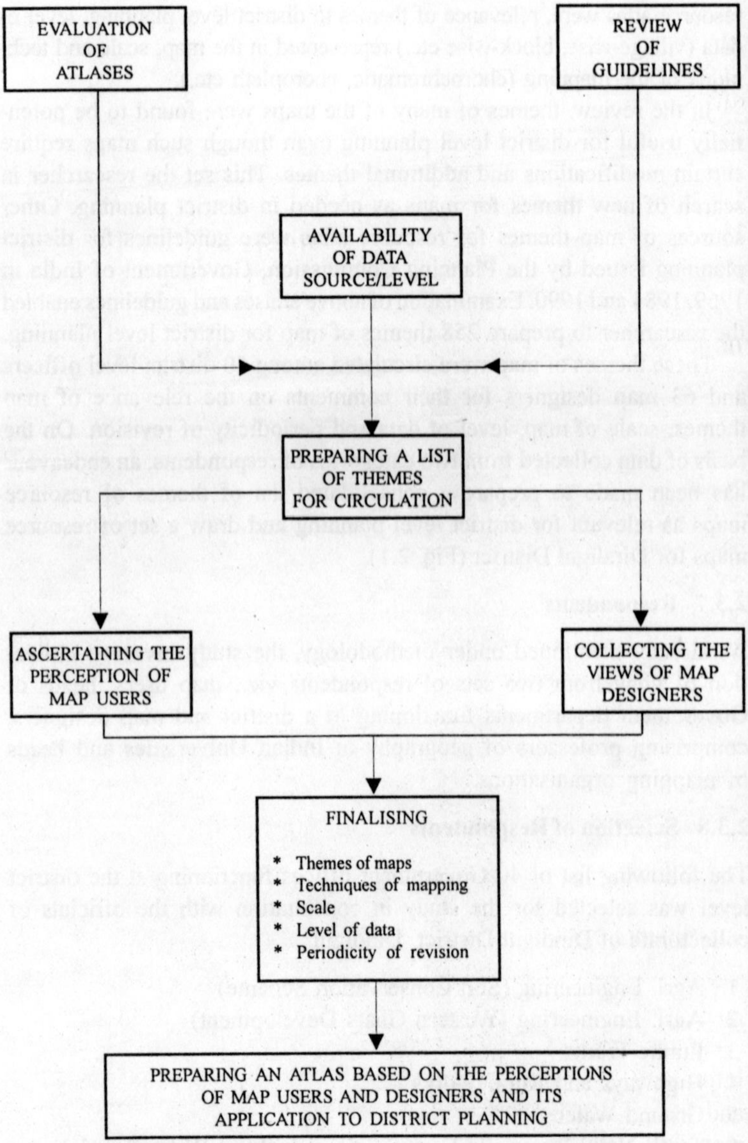

Fig. 2.1 : Methodology

9. Health Services
10. Survey and Land Records
11. District Forest Office
12. District Technical Cell
13. Joint Director of Agriculture
14. Asst. Director of Agriculture (Oilseeds)
15. Asst. Director of Agriculture (Training and Visit)
16. Joint Director (Sericulture)
17. Deputy Director of Horticulture
18. Joint Director of Animal Husbandry
19. Fisheries
20. Lead Bank (Canara Bank)
21. District Industrial Centre
22. Regional Transport Office
23. Joint Director of Cooperation
24. Deputy Director of Cooperation (Dairying)
25. Chief Education Officer
26. Non-formal and Adult Education Office
27. District Revenue Office
28. District Social Welfare Office
29. District Adi-Dravida Welfare Office
30. District Backward Class Welfare Office
31. Family Welfare
32. Child Development Project Office
33. District Rural Development Agency
34. District Panchayat Development Office
35. Tamil Nadu Electricity Board
36. District Town Panchayat Office
37. Tamil Nadu Civil Supplies Corporation
38. District Supply Office
39. Khadi and Village Industries Board
40. Superintendent of Police

The Fig. 2.1 users were contacted for the study by using census method.

A selected list of map designers was prepared on the basis of information available in the Hand Book of Common Wealth Universities and Indian Cartographer Journal. The following is the list of selected map designers.

1. Aligarh Muslim University

2. Andhra University
3. Banaras Hindu University
4. Bangalore University
5. Bhagalpur University
6. Bihar University
7. Doctor Harisingh Gour Vishwa Vidyalaya
8. Gauhati University
9. Gujarat University
10. Jamia Millia Islamia
11. Karnataka University
12. Kumaun University
13. Kurukshetra University
14. Madurai Kamaraj University
15. Magadh University
16. Maharshi Dayanand University
17. Mohanlal Sukhadia University
18. Mysore University
19. North Eastern Hill University
20. Osmania University
21. Patna University
22. Punjab University
23. Punjabi University
24. Rajasthan Vidyapeeth
25. Ranchi University
26. Ravi Shankar University
27. Shanti Niketan
28. Shivaji University
29. Sri Krishnadevaraya University (Head)
30. Sri Krishnadevaraya University (Professor)
31. Sri Venkateswara University
32. University of Allahabad
33. University of Baroda
34. University of Bombay
35. University of Burdwan
36. University of Calcutta
37. University of Delhi
38. University of Garhwal
39. University of Jammu
40. University of Jodhpur
41. University of Kashmir

42. University of Madras
43. University of Mysore
44. University of Poona
45. University of Rajasthan
46. University of West Bengal
47. Utkal University

In addition, heads of the following mapping organisations were contacted for the present study:

48. Research and Development, Survey of India
49. National Atlas and Thematic Mapping Organisation (NATMO), Director
50. Survey of India
51. National Remote Sensing Agency
52. NATMO, Deputy Director
53. National Institute of Rural Development
54. Automated Cartography Cell, Survey of India
55. Survey Training Institute, Survey of India
56. Institute of Remote Sensing
57. Centre for Studies in Resource Engineering
58. Centre for Earth Science Studies
59. Centre for Economic and Social Studies
60. Resource Analysis Division, Centre for Earth Science Studies
61. Centre for Study of Regional Development
62. Cartography Cell, Survey Training Institute, Survey of India
63. French Institute of Pondicherry.

Among these 63 map designers, only 19 of them sent the completed questionnaires despite three reminders.

2.3.9 Instruments for Data Collection

Primary data for this investigation relating to the map users (Government departments) were collected in two rounds using two pre-tested interview schedules. The first round of data collection covered all the 40 departments by using the schedule given in Appendix I.

In the second round of survey, the researcher distributed 258 maps themes, which includes 49 model resource maps (Appendix II) among the above 40 map users. The details of perception of map users, on matters like scale, level of data, colouring, additional maps required etc., were collected in another pre-tested schedule (Appendix III).

The list of 258 map themes and also another list of 75 themes, as required by the users along with a pre-tested questionnaire (Appendix IV) were sent to 63 map designers with a request to give their views about themes of resource maps, level of data, scale of maps and periodicity of revision for district planning.

2.3.10 Data Analysis and Presentation

Data collected from the map users in two pre-tested schedules were processed in the computer, after scrutinisation of the schedules for correctness and completeness. Further, secondary data on various sectors of development like agriculture, industry etc., collected in the first round for drawing model maps were processed manually and a total of 49 maps were drawn. In the second round, these maps were presented to the map users for ascertaining their perceptions. In the third round, the views from 19 map designers were gathered and processed manually.

In the analysis of data, suitable statistical techniques like – percentage, average, rate, proportions, quartiles were used. The study has used data for the period 1989-90 for the preparation of model resource maps as this study was undertaken during 1988-89.

The 49 maps which were circulated in the second round of data collection were revised in 34 Maps on the basis of the perception of map designers. Computer was used in redrawing some of the maps. The maps thus revised and redrawn form part of Chapter 4 of this book.

2.3.11 Outcome of the Study

1. The present exercise facilitated review of the data base and information systems, obtained at the district level and the extent of their utilisation.
2. A comprehensive list of map themes for district level planning has been prepared as a result of this work.
3. The study also involved preparation of appropriate resource maps on selected themes for district level planning with particular reference to Dindigul District.

2.3.12 Significance of the Study

Government departments functioning at the district level implement and also monitor development programmes without proper and adequate resource data. The present study assumes significance by supplying resource data in the appropriate form to the district level officials as perceived and required by them. The study also provides a comprehensive list of themes and techniques for the preparation of a district level resource atlas for planning purposes. Thus, it is hoped to pave way for scientific documentation of resource data for the present and future use by preparing a resource atlas as required by the officials.

2.3.13 Chapterisation

The study is presented in seven chapters. The first chapter deals with the introductory aspects of the study and the second chapter explains review of literature and methodology. The status of database and information systems available at the district level are studied in the third chapter. The details of evaluation of resource atlases are presented in the fourth chapter while the perception of map users and designers are discussed in the fifth chapter. The themes and techniques of proposed resource atlas are finalised and model resource maps on selected themes for Dindigual District are presented in the sixth chapter. Summary and Conclusion forms the seventh chapter of this book.

2.3.14 Limitations

1. There are many atlases available at national and state levels, due to paucity of time, the researcher could review only one national, two state level and one district level atlas.
2. The present exercise in perceptional cartography is limited to the perceptions of map suers and views of map designers on map themes, scale of map, level of data and periodicity of revision as required for district level planning.
3. Though 258 themes were circulated among the map users in the first round of data collection, only 49 model maps were prepared to motivate them to spell out with their perceptions. These maps were re-drawn into 34 Maps based upon the outcome of the analysis of the perceptions and requirements of the map user and designer.
4. The remote sensed data could not be used by the researcher for want of infrastructural facilities like computer, and software for image processing.

REFERENCES

1. Government of India, *National Atlas of India,* Hindi Edn., Ministry of Education and Scientific Research, Calcutta, 1957.
2. Government of India, *Indian Agricultural Atlas,* Ministry of Food and Agriculture, New Delhi, 1958.
3. Director of Economics and Statistics, *Economic and Statistical Atlas of Madhya Pradesh,* Bhopal, 1958.
4. Government of Mysore, *Mysore State in Maps,* Department of Statistics, Bangalore, 1958.
5. *Indian Oilseeds Atlas*, Indian Central Oilseeds Committee, Hyderabad, 1958.
6. *Atlas of Indian Lac,* Indian Lac Cess Committee, Ranchi, 1958.
7. *Coconut Atlas of India,* Indian Central Coconut Committee, Ernakulam, 1959.
8. *Indian Tobacco Atlas,* Indian Central Tobacco Committee, Madras, 1959.
9. *Indian Cotton Atlas,* Central Cotton Committee, Bombay, 1959.
10. *Indian Jute Atlas,* Indian Central Jute Committee, Calcutta, 1959.
11. *Agricultural Atlas of Madhya Pradesh*, Directorate of Land Records, Madhya Pradesh, Gwalior, 1959.
12. Government of Orissa, *Statistical Atlas of Orissa,* Bureau of Statistics and Economics, Bhubaneswar, 1959.
13. Government of Rajasthan, *Statistical Atlas of Rajasthan*, Directorate of Economics and Statistics, Jaipur, 1959.
14. Government of Andhra, *Statistical Atlas of Andhra State*, Bureau of Economics and Statistics, Hyderabad, 1959.
15. *Nutrition Atlas of India,* National Institute of Nutrition, Indian Council for Medical Research, Hyderabad, 1959.
16. *Diet Atlas of India*, National Institute of Nutrition, Indian Council for Medical Research, Hyderabad, 1959.
17. Government of India, *Atlas of Northern Frontier of India,* Ministry of External Affairs, New Delhi, 1960.
18. L.T.A. Learmonth and L.S. Bhat, *Mysore State: An Atlas of Resources,* Calcutta, 1961.
19. *Economic Atlas of Madras State,* National Council for Applied Economic Research, New Delhi, 1962.
20. S.L. Duggar, *Agricultural Atlas of Punjab*, Agricultural University, Hissar, 1966.
21. S.P. Chatterjee, *Planning Atlas of Damodar Valley Region,* Damodar Valley Corporation, Calcutta, 1969.
22. *Burdwan District Atlas*, Department of Development and Planning, Calcutta Metropolitan Planning Organisation, Calcutta, 1969.
23. A.K. Sen, *Agricultural Atlas of Rajasthan,* Indian Council of Agricultural Research, New Delhi, 1970.
24. *Census Atlas of India,* Census of India, New Delhi, 1970.
25. *Climatological Atlas of India,* Indian Meteorological Department, Pune, 1971.
26. *Rainfall Atlas of India,* Indian Meteorological Department, Pune, 1971.
27. *Administrative Atlas of Haryana,* Census of India, New Delhi, 1972.
28. *Administrative Atlas of Tamil Nadu*, Census of India, New Delhi, 1972.
29. *Administrative Atlas of Rajasthan,* Census of India, New Delhi, 1976.

30. *Administrative Atlas of Andhra Pradesh,* Census of India, New Delhi, 1977.
31. *Irrigation Atlas of India,* National Atlas and Thematic Mapping Organisation, Calcutta, 1972.
32. P.K. Tiwari, *Agricultural Atlas of Uttar Pradesh*, G.B. Pant University of Agriculture and Technology, Pant Nagar, 1973.
33. B.P. Ambasthya, *James Rennell's Memoir of a Map of Hindustan and the Bengal Atlas,* N.V. Publications, Patna, 1973.
34. M. Alam, *Planning Atlas of Andhra Pradesh,* Osmania University, Hyderabad, 1974.
35. J. Singh, *Agricultural Atlas of India,* Vishal Publication, Kurukshetra, 1974.
36. *Tourist Atlas of India,* National Atlas and Thematic Mapping Organisation, Calcutta, 1974.
37. Government of India, *Indian Agricultural Atlas,* Department of Agriculture, New Delhi, 1975.
38. Government of Pondicherry, *Statistical Atlas of Union Territory,* Bureau of Statistics and Valuation, Pondicherry, 1976.
39. *Atlas of Forest Resources of India,* National Atlas and Thematic Mapping Organisation, Calcutta, 1976.
40. Government of Kerala, *Kerala in Maps*, Bureau of Economics and Statistics, Trivandrum, 1978.
41. *Agro-Climatic Atlas of India*, Indian Meteorological Department, Pune, 1978.
42. *Indian Census Centenary Atlas*, Census of India, New Delhi, 1979.
43. *Indian Literacy Atlas,* Central Institute of Indian Languages, Mysore, 1978.
44. *Atlas of Agricultural Resources of India,* National Atlas and Thematic Mapping Organisation, Calcutta, 1980.
45. *National Atlas of India*, National Atlas and Thematic Mapping Organisation, Calcutta, 1982.
46. A. Ramesh and P.S. Tiwari, *Basic Resource Atlas of Tamil Nadu,* University of Madras, Madras, 1983.
47. *Resource Atlas of Kerala,* Centre for Earth Science Studies, Trivandrum, 1984.
48. *Social and Economic Atlas of India,* Oxford University Press, Madras, 1987.
49. *Planning Atlas of Medak District,* Centre for Economic and Social Studies, Hyderabad, 1988.
50. V.C. Joshi, Village Maps for Rural Development, *Indian Cartographer,* Vol. VIII, 1988, p. 64.
51. L.S. Bhat *et al.*, *Micro Level Planning — A Case Study of Karnal Area, Haryana, India,* K.B. Publications, New Delhi, 1976 pp. 9-15.
52. P.C. Tiwari, *Regional Development and Planning in India,* Criterion Publications, New Delhi, 1988, pp. 38-225.
53. V. Venugopal Reddy, *Multi Level Planning in India,* Vikas Publishing House, New Delhi, 1979, p. 155.
54. Government of India, Planning Commission, *Report of the Working Group on District Planning*, Vol. I, New Delhi, 1984, p. 32.
55. De Brommer Stephane, Proposal for Standardisation of Symbols on Thematic Maps, *International Year Book of Cartography,* Vol. 6, 1966, pp. 47-77.
56. M. Kamkov Anatolij, The International Language of Geographical Maps, *International Year Book of Cartography*, Vol.11, 1971, pp.209-213.
57. M.I.Nikishov and Preobrazhensky, The Problems of Unification of the Contents and Conventional Signs Standardisation on Maps. *International Year*

Book of Cartography, Vol.11, 1971, pp.127-135.

58. Ratjaski Lech, The Methodical Basis of the Standardisation of Signs on Economic Maps. *International Year Book of Cartography,* Vol. 11, 1971,. pp. 137-138.
59. J. Ross Mackay, Experiments with Some Symbols and Map Projections, *Annals of the Association of American Geographers,* Vol. 44, No. 2, 1954, pp. 225-226.
60. I. Clarke John, Statistical Map Reading, *Geography,* Vol. 44, No. 2, 1954, pp. 96-104.
61. Ekman Goesta and Junge Kenneth, Psychological Relations in Visual Perception of Length, Area and Volume, *Scandinavian Journal of Psychology,* Vol. 2, No.1, 1961, pp. 1-10.
62. J. Flannery James, The Relative Effectiveness of Some Common Graduated Point Symbols in Presentation of Quantitative Data, *Canadian Cartographer,* Vol. 8, No. 2, 1971, pp. 96-109.
63. Joachim Meinhoefer Hans, The Utility of the Circle as an Effective Cartographic Symbol, *Canadian Cartographer,* Vol. 6, No. 2, 1969, pp.105-117.
64. ________________, The Visual Perception of the Circle in Thematic Maps/ Experimental Results, *Canadian Cartographer,* Vol. 10, No. 1, 1973, pp. 63-84.
65. V. Crawford Paul, Perception of Graduated Squares as Cartographic Symbols, *Cartographic Journal,* Vol. 10, No. 2, 1972, pp. 85-88.
66. J. Phillips Richard and Noyes Liza, An Investigation of Visual Clutter in the Topographic Base of a Geological Map, *Cartographic Journal,* Vol. 19, No. 2, 1982, pp. 122-132.
67. Mark Shurtleff and R.Edward Geiselman, A Human Performance Based Evaluation of Topographic Maps and Map Symbols with Novice Map Users, *Cartographic Journal,* Vol. 23, No.1, 1986, pp. 52-55.
68. F.A. Taylor, *Colour Technology,* Oxford University Press, London, 1962.
69. J. Cuff David, Colour on Temperature Maps, *Cartographic Journal,* Vol. 10, No.1, 1973, pp. 17-21.
70. C. Patton Jeffrey and V. Crawford Paul, The Perception of Hypsometric Colours, *Cartographic Journal,* Vol. 14, No. 2, 1977, pp.115-127.
71. Barbara S. Bartz, Experimental use of Search Task in an Analysis of type Legibility in Cartography, *Cartographic Journal,* Vol. 7, No. 2, 1970, pp. 103-112.
72. R.V. Gerber, An International Study of Children's Perception and Understanding of Type used on Atlas Maps, *Cartographic Journal,* Vol. 19, No. 2, 1982, pp. 115-121.
73. R.M. Taylor, An Application of Subjective Scaling Techniques to Map Evaluation, *Cartographic Journal,* Vol. 11, No. 2, 1974, pp. 74-81.
74. J. Phillips Richard, Alan De Lucia and Nicholas Skelton, Some Objective Tests of the Legibility of Relief Maps, *Cartographic Journal,* Vol. 12, No. 1, 1975, pp. 39-46.
75. Chang Kang Tsung, Visual Aspects of Class Intervals in Choroplethic Mapping, *Cartographic Journal,* Vol. 15, No. 1, 1978, pp.42-48.
76. L.William Robert, Map Symbols: Equal Appearing Intervals for Printed Screens, *Annals of the Association of American Geographers,* Vol. 48, No. 2, 1958, pp. 132-139.
77. F. Jenks George and S. Knos Duane, The Use of Shading Patterns in Graded

Series, *Annals of the Association of American Geographers*, Vol. 51, No. 3, 1961, pp. 316-334.

78. C. Strossel Otto, *Standard Printing Screen System*, St. Louis, USAF Aeronautical Chart and Information Centre, 1972.
79. J. Cuff David, Shading on Choropleth Maps : Some Suspicions Confirmed, *Proceedings of the Association of American Geographers*, Vol. 5, 1973, pp. 50-54.
80. Clande Muller Jean, Association in Choropleth Map Comparison, *Annals of the Association of the American Geographers*, Vol. 65, No. 3, 1975, pp. 403-413.
81. S. Monmonier Mark, Class Intervals to Enhance the Visual Correlation of Choropleth Maps, *Canadian Cartographer*, Vol. 12, No. 2, 1975, pp. 161-178.
82. F. Jenks George and C. Caspall Fred, Varieties Exaggeration in Three-Dimensional Mapping, *Technical Paper*, No. 2, NR.389-146, Nonr 583 (15), Geography Branch, Office of the Naval Research, 1967.
83. Alan De Lucia, The Effect of Shaded Relief on Map Information, *Cartographic Journal*, Vol. 9, No. 1, 1972, pp. 14-18.
84. C. Grant Head, Land-Water Differentiation in Black and White Cartography, *Canadian Cartographer*, Vol. 9, No. 1, 1972, pp. 25-38.
85. J. Phillips Richard, Alan De Lucia and Skelton Nicholas, Some Objective Tests of the Legibility of Relief Maps, *Op. Cit.*, pp. 39-46.
86. F. Jenks George and C.Caspall Fred, Error on Choropleth Maps: Definition, Measurement, Reduction, *Annals of Association of American Geographer*, Vol. 61, No. 2, 1971, pp. 217-244.
87. Brandes Donald, Relative Effectiveness of Selected Relief Representation Methods for Topographic Maps, Unpublished *Master's thesis*, University of Calgary, Alberta, Canada, 1975.
88. Alan A. De Lucia and Donald W. Hiller, Natural Legend Design for Thematic Maps, *Cartographic Journal*, Vol. 19, No. 1, 1982, pp. 46-52.
89. D. Sheppard and J.M. Adams, A Survey of Driver's Opinion on Maps for Route Finding, *Cartographic Journal*, Vol. 8, No. 2, 1971, pp. 105-114.
90. G. McGrath and H.W. Castner, The Design of a Tourist Map Series Final Report under Contact O2SV, 23244-5-5596, Department of Energy, Mines and Resources, Surveys and Mapping Branch, Topographical Survey Directorate.
91. Sorrel Patrick, Map Design — With the Young in Mind, *Cartographic Journal*, Vol. 11, No. 2, 1974, pp. 82-90.
92. T.K. Gurusamy, *Annual Report of INCA Commission of School Atlas*, IV National Conference, Madras, 1982, pp. 1-7.
93. S.K. Sahni, To Re-design and Construct Topographic Map on Scale 1:25000 — A suggestion, IV Indian National Cartographic Association Conference, Madras, 1982, pp. 1-10.

3

A STUDY ON DATA BASE AND INFORMATION SYSTEM FOR DISTRICT LEVEL PLANNING

3.1 INTRODUCTION

District is the basic administrative unit in India. Having shifted the focus of planning from macro to micro level through decentralisation, it has become necessary to take steps to improve the data base for planning at the district level[1]. Officials working at the district level, while making decisions, have to refer to a variety of relevant data. They make use of both primary and secondary data in the decision making process. Quite naturally, the soundness of the decision depends on quality and availability of data used by them. The data base which are available for district level planning has some inherent problems.

3.2 PROBLEMS OF DISTRICT LEVEL DATA BASE

3.2.1 Compatibility

Major part of the data, derived from different sources, are collected for different purposes and thus they are incompatible. Lacking consistency, they are often unsuitable for sharing and re-using and there are also intractable problems of classification and codification of data and even legibility[2].

The Table 3.1 gives details about the respondents (district officials) and levels of data collection.

The table shows that the data are being collected at panchayat, revenue village, block, *taluk*, district and other levels. Of the total respondents, 44.7 per cent of the respondents collect data at the block level while 28.9 per cent of them at the revenue village level. Further, 10.6 per cent respondents collect data at other levels *viz.*, beat-wise,

Table 3.1 : Levels of Primary Data Collected by the Respondents

Sl. No.	Levels of Primary Data	Respondents in percentage
1.	*Panchayat*	10.6
2.	Revenue Village	28.9
3.	Block	44.7
4.	*Taluk*	2.6
5.	District	2.6
6.	Others (Beat/user/JE office/ Police Station)	10.6
	Total	100.0

user-wise, (Junior Engineer) office-wise or police station wise. Similarly, another 10.6 per cent of them collect data at panchayat level. From the above analysis, it is clear that about 45 per cent of them collect data at the block level, another 40 per cent collect at the revenue village and village panchayat level.

3.2.2 Time-lag

In the present system of administration, all the data required for planning, flow from districts to state governments and then to central government, after their consolidation first at the district level and then at the state level. This process has inherent delays and the experience shows that relevant data are rarely available when it is required for decision making/planning. Moreover, there are drawbacks with regard to completeness and accuracy[3].

There is time-lag between collection of data and their publication. As a result, in most of the cases the data are outdated at the time of publication[4]. The data for the current year is hardly available. For instance, when the researcher collected data for preparing model maps in March 1991, less than 20 per cent of the data related to 1989-90 and the rest to earlier years. About 18 per cent of the data was more than five years old.

3.2.3 Data Collection Personnel

Responsibility for collection of data often rests with people who are at the bottom of hierarchy, who are neither provided with appropriate training nor monetary incentives. Further, many of the officials like Assistant Agricultural Officers, Village Administrative Officers, and Rural Welfare Officers are saddled with a variety of responsibilities with meagre material resources and a few subordinate officials. These officials have to collect data from large areas like a revenue village or a cluster of panchayats.

Consequently, the quality of data collected by them remains diluted[5].

3.2.4 Non-availability of Data at One Place

Data on all the inter-related items are seldom available at one common place to various users[6]. The researcher too, in course of discussion with district officials, found them relying upon a number of sources for their data. The particulars of secondary data users and number of sources of data tapped by them are presented in the Table 3.2.

Table 3.2 : Distribution of Secondary Data Users According to Number of Sources of Data

Sl. No.	Number of sources of Data	Percentage of Secondary Data using Respondents
1.	1-2	37.9
2.	3-4	44.8
3.	5 and above	17.3
	Total	100.0

It is clear from the above table that about 63 per cent of respondents depend on three or more sources of data and about 17 per cent of them rely on five or more sources. It is evident from the above analysis that majority of the respondents collect data from three or more primary sources. They have to collect data by visiting a number of offices as the required data is not available at one source.

3.2.5 Data Integrity

The current method of data collection from different sources needs to be verified at the source and validated on an overall basis. Some amount of counter verification is required because of the use of different types of technologies of data gathering by different government departments, which do not always work in a coordinated manner. There are often multiplicity of agencies functioning under different government departments working in the area. Necessary coordination among them is not found and in consequence, duplication creeps-in, which means wastage of men and material resources of the government departments[7].

The data collected by a government department are frequently used by other departments for making decisions. The reliability of the data used by them decides the soundness of the decision. Hence, the need to ascertain the reliability of data collected by the primary data collecting personnel.

Investigation by the researcher reveals that about 65 per cent of primary data users consider the data collected by themselves as 'fully' reliable. Only 10 per cent of them rate the reliability of their data as 'moderately' reliable, while the remaining 25 per cent rank their data as 'largely' reliable. It is thus found, about two-thirds of the data are fully reliable.

3.2.6 User Need

The existing data bases are created only for the primary user-department and without considering the needs of other secondary users[8]. A lot of data have been collected, tabulated and published. From the user's view, the organisation of data is not appropriate at present and one has to do complex exercises to make out his portion of information[9].

3.3 NATIONAL INFORMATICS CENTRE

The importance of data source for district level administration can be appreciated from the fact that the departments prepare more than 700 reports (monthly, quarterly, half-yearly and annual) which are sent to various state/central Government departments. Naturally, measures have been underway to ensure timely collection, compilation, dissemination and on-line accessibility of information on several sectors of economy at the district, *taluk*, block, *panchayat* and village levels. In this regard, the work of National Informatics Centre (NIC) is significant.

NIC was conceived in 1973 by Government of India which nucleated it in 1976, as a high priority plan project under the Electronics Commission/Department of Electronics. NIC, has been designed to play a catalytic role in creating computer awareness and promoting system approach in data collection, organisation, processing and its on line accessibility. It started extending its computer based information service to State Governments and District administrations since 1985-86. In 1988, NIC was transferred from the Department of Electronics to the Planning Commission.

3.3.1 Objectives

The major objectives of the NIC are:

1. Development of computer-based information systems in various sectors of the economy for planning, project monitoring

and decision support,
2. Promotion of informatics culture at the district, state and national levels,
3. Improvement of the analytical capacity and presentation of statistics utilised for national, state and district planning,
4. Development of modelling, forecasting and simulation techniques that are required for planning and decision support,
5. Establishment of a computer network with necessary comunication equipment for ensuring easy access of information across the nation and ensuring reliable, fast and optimal use of resources,
6. Evolving of standards for data collection, compilation and dissemination, with the cooperation of related agencies at the district, state and central levels,
7. Creating a dedicated training school with branches in all parts of the country for extensive training/retraining of personnel in public services and users of NIC, and
8. Providing gateways for accessing International Data-base for those working in districts/states/centre[10].

3.3.2 Facilities

NIC has set up four regional entres at Delhi, Pune, Bhuvaneshwar and Hyderabad, each having main frame Super Computer (NEC S-1000 Model). Super Mini-Computers (ND-550 or equivalent) are installed at State capitals and Super PC/AT computers in the districts. NIC has established a nation-wide satellite (INSAT-1D) based Computer Communication Network called NICNET, to provide 'Informatics Services' to Government at different levels[11].

3.3.3 Use of Facilities

In Tamil Nadu, NIC provides following services at the State and District level.

1. Land records information system
2. Health information system
3. Land acquisition system
4. Agriculture census
5. DISPLAN data collection and processing
6. Monitoring price system of public distribution system items
7. Purchase and sales monitoring system
8. Monitoring system for performance of District Industries Centre

9. Paddy procurement
10. Control orders of civil supplies
11. Rainfall monitoring system
12. Crop forecasting analysis
13. Price data of commodities for Eco-stat
14. Camp-wise information of Sri Lankan Refugees
15. Small savings monitoring system
16. Stocks and issue particulars of TNCSC godown
17. Twenty Point Programme
18. Reconciliation of the treasury figures with AG's under old age pension scheme
19. DISNIC-AGRIS
20. Drought relief operation
21. Schedules of revenue department
22. NICMAIL services[12]

However, the researcher found that the above services/facilities had not been fully developed due to reluctance of the departments of state government to place all their data with NIC. These departments have supplied very limited volume of data to NIC on the condition that these particulars should not be supplied to others not even to other government departments without their consent. Therefore, NIC has yet to go a long in operating as a true information centre for all the bonafide users.

3.3.4 District Information Systems of National Informatics Centre (DISNIC)

NIC has launched in 1987 a programme to develop statistical information system and data bases at district level viz., District Information System (DISNIC)[13].

3.3.5 Objectives

The specific objective is the development of Informatics for the district administration in a coordinated manner using NICNET facilities.

The DISNIC has been implemented to:

1. develop the necessary informaion/data bases (both spatial and non-spatial) in various sectors of the economy for planning and decision making at the district level,
2. promote informatics culture at the district level through appropriate training programme,
3. facilitate information standardisation and discipline wherever

required,

4. improve the analysis capacity and the presentation of the statistics utilised for national, regional and district planning,
5. develop modelling and forecasting techniques that are rquired for decision making for socio-economic development, and
6. promote Geographic Information System (GIS) techniques for planning and also remote sensed data application for natural resources management at grass-root level[14].

In the initial stages, DISNIC was started in selected districts of Kerala, Karnataka, Orissa and Uttar Pradesh States as a pilot project and now it is operational in about 450 districts throughout the country[15].

Using the NICNET facilities at the district level, the information flow will get rationalised both upwards and downwards, DISNIC will thus pave way for easy collection, compilation and dissemination and on-line accessibility of information on several sectors of the economy at district level with the availability of information at all possible levels *viz*., 1. District, 2. *Taluk*, 3. Block, 4. *Panchayat,* and 5. Village[16].

3.3.6 Components of DISNIC

The following are the functional components of DISNIC (Fig 3.1).

(a) Management Information System (MIS) for the revenue administration,
(b) MIS for the development administration (sectoral data bases),
(c) District Planning Information System (DISPLAN),
(d) Geographic Information System of NIC (GISNIC), and
(e) National Natural Resource Management System (NNRMS)[17].

(a) MIS for Administration at the District Level: MIS will be created in the following sectors for revenue administration.

* Revenue Recovery — Land revenue recovery
  — Miscellaneous revenue recovery
* Land Administration — Land records management
  — Land reforms
  — Land acquistion
  — Land assignment
* Implementation of housing schemes
* Functional literacy programme
* Relief works following natural calamity
* Law and order

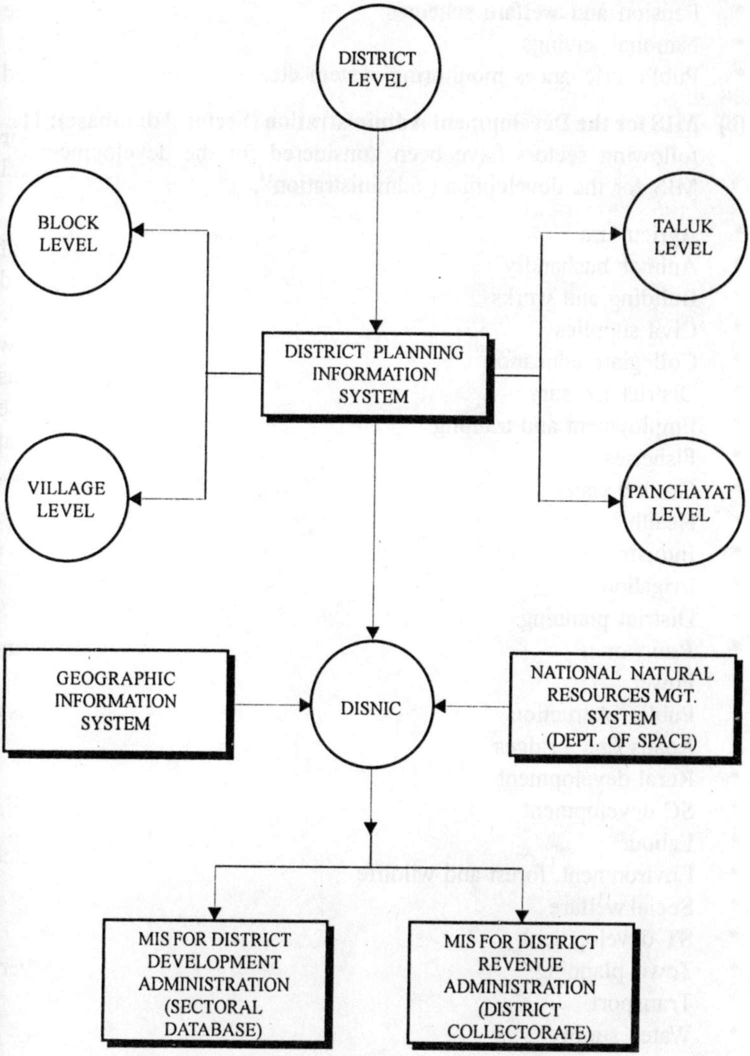

Source : National Informatics Centre, District Information System of National Informatics Centre, DISNIC Handbook, Government of India, Planning Commission, New Delhi, 1992, p. 20.

Fig. 3.1 : Functional Components of DISNIC

* Vigilance cases
* Pension and welfare schemes
* National savings
* Public grievances monitoring system etc.

(b) **MIS for the Development Administration (Sectoral database):** The following sectors have been considered for the development of MIS for the development administration[18].

* Agriculture
* Animal husbandry
* Building and works
* Civil supplies
* Collegiate education
* District treasury
* Employment and training
* Fisheries
* Ground water
* Health
* Industry
* Irrigation
* District planning
* *Panchayat*
* Power
* Public instruction
* Roads and bridges
* Rural development
* SC development
* Labour
* Environment, forest and wildlife
* Social welfare
* ST development
* Town planning
* Transport
* Water supply

(c) **DISPLAN**: Preparation of a comprehensive plan for a village, cluster of villages or district requires a strong spatially integrated socio-economic cum natural resources data-base for planning, management and decision making. DISNIC will facilitate planning at the district level.

Functional component of DISPLAN, aims at

1. Creation and maintenance of socio-economic, agro-economic and natural resource data bases.
2. Development of spatial decision support system, analytical tools and expert systems tools.

It was suggested to have a five level data base to store and retrieve the information[19] at all possible hierarchy of the district administration because of the corresponding nature of work at these levels viz., district, taluk, local body and village/ward levels (Fig 3.2).

Information under DISPLAN: The information under DISPLAN can be broadly classified into:

1. General parameters (those parameters which are not planned for but have to be taken into consideration during formulation of plan)

   General data base will contain the following:

   1. At a glance
   2. Geographical details
   3. Climate
   4. Population
   5. Land-use pattern

2. Sector-wise parameters (those which are needed for planning)

   It is proposed to create sector-wise data bases in the following sectors:

   1. Health
   2. Excise
   3. Local bodies
   4. Agriculture and irrigation
   5. Education
   6. Sales tax
   7. Water supply
   8. Transport and communication
   9. Housing
   10. Price
   11. Sanitation
   12. Flood control and anti-sea erosion
   13. Industry
   14. Judicial and administrative statistics

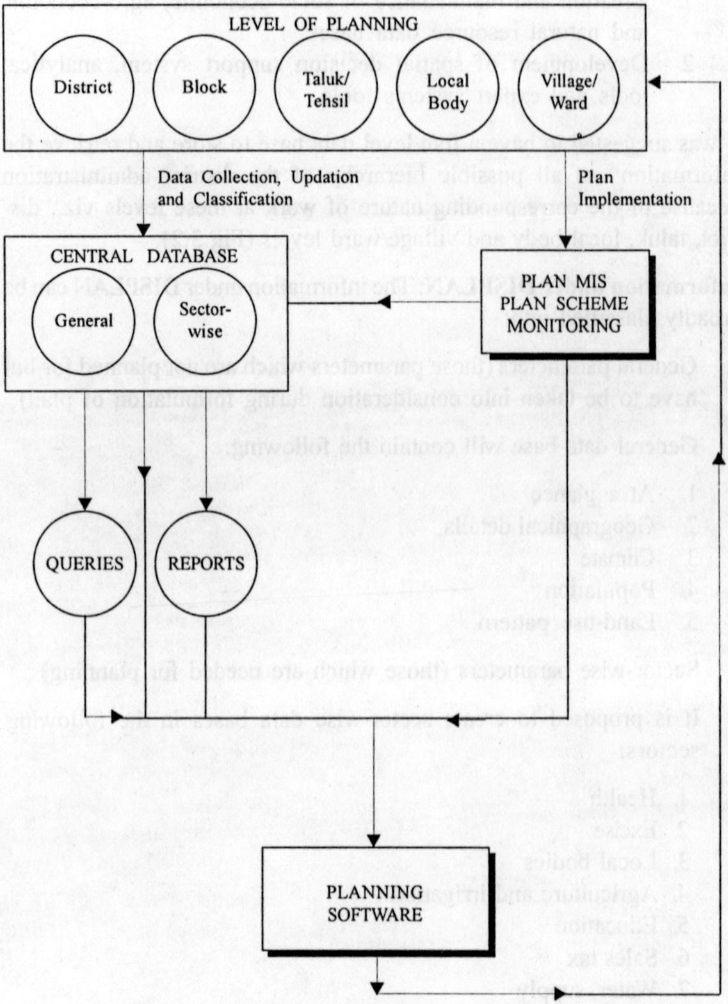

Source : Madaswamy Moni and N. Vijayaditya, DISNIC, A NICNET based District Government Informatics Programme in India, a paper presented at Indian Computing Congress, December, 1991, New Delhi.

Fig. 3.2 : District Planning Information System

15. Fisheries
16. Land reforms
17. Environment
18. Rural development schemes
19. Forestry
20. Cooperative scoieties
21. Institutional finance and banking
22. Employment and training
23. Power
24. Registration
25. Recreation
26. Minerals
27. Animal husbandry
28. Drainage and sewage etc.[20]

(d) **GISNIC**: Computer cartography is one of the components of Geographic Information System (GIS). GIS is a system designed to store, manipulate and display the geo-referenced or geo-coded data, capable of handling, analysing and modelling both locational data and attributes or descriptive data about features[21]. Maps are a natural component of GIS, as they organise spatial information for convenient use. The use of computers in cartography has not only increased the productivity of a cartographer but also indicated the new areas, where maps could be useful.

The present GISNIC aims at a map based decision support system or computer-based cartographic system. The implementation of this system will lead to an omnipotent GIS with the capabilities for handling remotely sensed data and user friendly expert shells[22]. GIS software using state of art concepts were developed to facilitate the development of GIS for enabling spatial planning. The urban GIS project covers metro cities. For storage and retrieval of attribute data, various data bases developed under DISNIC will be used.

(e) **National Natural Resource Management System**: NIC has plans to set up NICNET facilities in all the regional remote sensing centres of NNRMS to use it for further analysis at the district level by the NIC centres.

Though DISNIC was launched simultaneously in all the districts, there has been a gradual and phased implementation in the introduction of its above 5 functional components. For instance,

NIC in Dindigul District, Tamil Nadu has introduced DISPLAN only, under DISNIC. And as part of this project they have been feeding data on education, agriculture and village profiles.

3.3.7 Special Projects under NIC

(a) Tele-informatics Development and Promotion Programme (TDPP): It is a UNDP assisted programme to develop technology for mass information dissemination with demonstration of its application in operational system. Its PC based tele-text de-coder that can receive television signals containing tele-text data. The data are stored on disc and retrieved on PC screen.

(b) PC based Magazines and Page Assembler and Distributor (MPAD) — converts raw data obtained from computer or manual sources into tele-text magazines and thus allowing connection of more types of information sources to the tele-text system.

(c) General Information Service Terminal of NIC (GISTNIC) : NIC has launched 'dial-up for data service' with an aim to make general information available throughout the country on NICNET.

The objectives of this GISTNIC are :

1. to offer on-line general information/data base services to common public by deriving useful information from various existing data bases and other information sources,
2. to act as a NICNET based communication channel for the government to assist in disseminating important development information to common public, and
3. to popularise information culture into the life-style of common public for improving quality of life.

The GISTNIC is already operational through NICNET or at GISTNIC booths installed at convenient places like Government secretaries, State Legislative Assemblies, libraries, airports, hotels, etc.

It offers on-line information services on the following :

1. Statistical profile of India
2. District level economic indicators

3. District profiles
4. University education guide of India (course-wise)
5. Tourist guide of India
6. All India railway time-table
7. Hotel guide of India
8. General information guide
9. All India hospital directory
10. Country profile
11. World economy tables
12. Industrial directory of India (unit-wise)
13. Employment guide, and
14. Investment guide

There are some specialised data bases built, based on the data supplied from international/national organisations – the World Bank, the International Monetary Fund and the Census Commissioner/Registrar General of India (Village-wise Census data comprise about 100 parameters, village level details on infrastructure and service facilities and people related information. *Taluk* level and district level aggregates are also available on the above parameters.)[23]

However, provisions of these facilities in the above centres have not been sufficiently publicised and therefore, very few are aware of the availability of these services.

(d) NIC-ICMR Bio-medical Information Centre set up in 1987, has been providing operational services through the international bibliographic data base for medicine, called Medlars on NICNET throughtout the country. The Union catalogue for scientific serial compiled by INSDOC has been updated on computers by the centre using the NEC S-1000 computer. NIC provides information on international patents from its patent information system which is regularly updated by procuring tapes from the international data bases.

(e) The main objective of NICNET, 'Project Monitoring Programmes' is to provide project monitoring facilities to the Project Monitoring Committee of the public sector undertakings to enable them to monitor the large public sector projects. Currently, the Steel Authority of India Ltd., (SAIL), National Thermal Power Corporation (NTPC) and Food Corporation of India (FCI) are using the NICNET facilities. Now the facility is extended to other public sector undertakings.

3.3.8 Use of NIC in future

NIC will in course of time be able to provide computer based services to Central Government departments, State Government Departments and District Administration. The ND-550 machine installed at state capitals will be linked to NICNET with modified software. Work on Information System will be taken up for various central government departments.

Micro modelling project will be initiated at the District level to facilitate micro level planning. The Grass-Root level Information Development (GRID) Programme will be implemented in a few pilot districts for providing computer based services at Block/Tehsil level. GISNIC software will be made available at the District nodes for regional planning. The State centres will provide computer based service to the state government departments on an increased basis[24].

However, the said development will be possible, only when the department at district level shed their inhibition in parting with their data, cast away the veil of secrecy which they have unnecessarily placed on their own fund of information.

3.3.9 Other Computerised Information Systems

Besides NIC, there are other computerised information system available. Important systems among them are Computerised Rural Information System Project (GRISP), National Resource Data Management System (NRDMS), Technology Information, Forecasting and Assessment Council (TIFAC) and a computerised country-wide data transmission network of Computer Maintenance Corporation (CMC).

(a) GRISP: Initially GRISP was introduced in 1986 on a pilot basis in Chittoor in Andhra Pradesh, Hazaribagh in Bihar, Quilon in Kerala, Betul in Madhya Pradesh, Thome in Maharashtra, Cuttack in Orissa, Patiala in Pubjab, Udaipur in Rajasthan, Kanpur in Uttar Pradesh and 24-Parganas in West Bengal. After a detailed review of its functioning, CRISP was extended to District Rural Development Agencies in 422 districts in 1987[25] and now this facility is available in all districts in India. This programme facilitates easy handling of the massive data and information of rural development project at the beneficiary as well as village level[26].

(b) National Resource Data Management System (NRDMS): It was started in Visakhapatnam in Andhra Pradesh, Sultanpuri in Uttar Pradesh, Gurgaon in Haryana, Koraput in Orissa, Khera in Gujarat and Manipur, as a pilot project of Department of Science and Technology. These projects are planning to make out blue prints for development at various levels — village, block and district – utilizing data available from the NRDMS, Census, National Sample Survey etc. It seeks to establish linkages with the programmes of the National Natural Resource Management System (NNRMS), NICNET and other sectoral information systems[27]. It endeavours to establish an area specific data management system for decentralised planning.

(c) INDONET: Another country-wide data transmission network known as 'INDONET' was started by the CMC in 1984. It has nodes in about 10 cities and uses leased lines from the Department of Telecommunication. It is used by customers like Welcome Group of Hotels and Madura Coats. It is still not in the picture since it lacks data base and relevant software[28].

(d) TIFACLINE: It is a nationally accessible information system of Technology Information Forecasting and Assessment Council (TIFAC) which was set up in 1986, under the Department of Science and Technology. TIFACLINE of TIFAC aims at providing technology information which will serve forecasting and assessment of needs of entrepreneurs, decision makers and planners.
The Indian users are linked with other world wide data bases with the installation of the Gateway Packet Switching System (GPSS) in Bombay by the Videsh Samachar Nigam Ltd. (VSNL). This international packet switch is connected to mini packet switches called Packet Assemblers-Disassemblers (PADS) in Bangalore, Delhi, Pune and Hyderabad which can be accessed by customers in those cities through the regular telephone lines.
NRDMS and NNRMS have already been linked to help district planning. NICNET too has a plan to set up its Centres at the centres of NNRMS for using the facilities of NNRMS for district level planning.

3.4 NICNET and Users

Reference has already been made to the reluctance of the departments at district level to store their data at NIC because of their unwillingness to place them for public perusal. It is true that the computer system of NIC has many safeguards like pass word, file protection against misuse of data. Yet, many of the departments do not repose confidence in the safeguards and they consequently wary of using these facilities. It is found that only 30 per cent of the departments at district level in the study area use the services of NIC in a limited way.

3.4.1 Areas of Use

The NICNET facilities are used by the departments functioning at District level. The uses are tabulated in the Table 3.3.

Table 3.3 : Areas of Use of NICNET at District Level

Sl. No.	User Department	Generation of Reports	Data Retrieval	Drawing Graphs/ Diagrams	Data Transfer
1.	Revenue	x	x	-	-
2.	Agriculture	x	x	-	-
3.	Lead/Coop. Bank	x	x	-	-
4.	Supply Office	x	x	x	x
5.	Panchayat Development	x	x	-	-
6.	Education	x	x	x	x
7.	Health	x	x	x	-
8.	District Industries Centre	x	x	-	-
9.	TWAD	x	x	x	-
10.	Animal Husbandry	x	x	-	-
11.	TINP	x	x	-	-
12.	Statistics	x	x	-	-
	Total in percentage	100	100	25	17

x Users, — Non Users

From the Table 3.3, it will be found that all NICNET users are employing NIC computer for reporting as well as data retrieval. One-fourth of them use it for drawing graphs/diagrams while only 17 per cent use it for transferring data to state level offices.

It is evident from the above analysis that the NICNET facilities are used for conventional uses like report generation and retrieval of data, only a small group of them for visual representation and data transfer.

From the above account, it will be evident that NICNET, CRISP, NRDMS, TIFACLINE etc., represent a significant advance in collection, compilation, dissemination of information on various sectors of economy at district and lower levels. However, these developments are yet to be fully astride eventhough some of them are about six years old. There are three main reasons for tardy development in this regard.

Insufficiency of cooperation between 'NIC' an organisation of

Central Government and district level offices, most of which belonging to State Governments have hamstrung progress. The district level offices are reluctant to use these new facilities and in the study area, users of these new facilities form only 30 per cent of the district offices. Even those offices which use them do not allow NIC to share their data with other departments. As a result the role of NIC at present is inadequate.

Secondly, multipicity of levels of data-revenue village/panchayat level-make it difficult for NIC to organise their data in a common frame for use by others.

Thirdly, access to remote-sensed data is yet to be available to NIC in Tamil Nadu.

At present the data base available in the NIC, located in Dindigul District is therefore incomplete. And they are not accessible to the users because of the restriction imposed by the State Government Departments. The only consoling feature in NIC is the functioning of GISTNIC since February, 1993 supplying Census data-1991 upto village level and other general purpose data like air/rail timings, hotels etc. Hence, the available data bases with NIC have not been used by the researcher in his work.

REFERENCES

1. National Informatics Centre, *District Information System of National Informatics Centre, DISNIC* - DISNIC Handbook, Government of India, Planning Commission, New Delhi, 1992, p. 19.
2. Ministry of Environment & Forests, INDO-US Workshop on Geographic Information System, *Background Papers,* Government of India, New Delhi, 1989, p. 2.
3. National Informatics Centre, *District Information System of National Informatics Centre, DISNIC*, DISNIC Handbook, *Op.cit.*, p. 59.
4. Ministry of Environment & Forests, INDO-US Workshop on Geographic Information System, *Background Papers, Op.cit.*, p. 2.
5. Utpal K. Banerjee, Key Considerations in Data Collection, *The Hindu,* June 22, 1989, p.18.
6. National Informatics Centre, *District Information System of National Informatics Centre, DISNIC* — DISNIC Handbook, *Op.cit.*, p. 11.
7. Utpal K. Banerjee, Key Consideration in Data Collection, *The Hindu, Op.cit.*, p. 18.
8. *Ibid.*, p.18.
9. National Informatics Centre, *District Information System of National Informatics Centre-DISNIC*, DISNIC Handbook, *Op.cit.*, p. 59.
10. National Informatics Centre, *District Information System of National Informatics Centre - DISNIC,* DISNIC Handbook, *Op.cit.*, p. 11.
11. National Informatics Centre, Performance Budget — 1991-92, Government of India, Ministry of Planning, Planning Commission, New Delhi, 1991,

pp. 1-2.

12. National Informatics Centre, Performance Budget — 1992-93, Government of India, Ministry of Planning, Planning Commission, New Delhi, 1992, pp. 16-17.
13. Madaśwamy Moni and N.Vijayaditya, DISNIC, A NICNET based District Government Informatics Programme in India, *a paper* presented at Indian Computing Congress, December, 1991, p.12.
14. Idem.
15. *Ibid,* p. 22.
16. National Informatics Centre, *District Information System of National Informatics Centre,* DISNIC - DISNIC Handbook, *Op.cit.,* p. 20.
17. *Ibid.*, pp. 20-21.
18. *Ibid.*, pp. 22-23.
19. *Ibid.*, p. 23.
20. *Ibid.*, pp. 68-71.
21. *Ibid.*, p. 24.
22. National Informatics Centre, Performance Budget 1992-93, *Op.cit.*, p. 5.
23. National Informatics Centre, General Information Service Terminal, Government of India, Planning Commission, New Delhi, 1992, pp. 1-5.
24. National Informatics Centre, Performance Budget 1991-92, Government of India, *Op.cit.*, p. 8.
25. Madaswamy Moin and N. Vijayaditya, *Op.cit.*, p. 9.
26. Utpal K. Banerjee, Information Gathering from District level, *The Hindu,* June 8, 1989, p.18.
27. *Ibid.*, p. 18.
28. N.N. Satchidanand, India Entering Information on top era, *The Hindu,* 4th July, 1990, p. 8.

4

AN ANALYSIS OF THEMES AND TECHNIQUES FOR DISTRICT RESOURCE ATLAS

4.1 INTRODUCTION

In India, a large number of atlases are available for various purposes at national and state levels. However, there are very few resource atlases which may help planning at district level. Any planning exercise has to depend upon the evaluation of available resources, their locations and pattern of their existing utilization. Hence the need for resource maps.

A study of the existing resource atlases not withstanding their unsuitability to the decentralised planning, can still enable a researcher to identify the themes, level of data, scale of map and techniques of presentation, as required for the decentralised planning and thus aid him to evolve an appropriate resource atlas for district level planning. Therefore, an attempt is made to evaluate the resource atlases which are now available. At the national level, in the absence of a comprehensive resource atlas, the only available atlas 'National Atlas of India' has been taken for evaluation. At state level, three comprehensive resource atlases are available – one each for Andhra Pradesh, Tamil Nadu and Kerala. Of these, Planning Atlas of Andhra Pradesh is the oldest atlas, published in 1974. Basic Resource Atlas of Tamil Nadu and Resource Atlas of Kerala were brought out in 1982 and 1984 respectively and these two latest atlases have been chosen for review at the state level. At the district level, the only available 'Planning Atlas of Medak District' has been selected for evaluation. For the purpose of arriving at a list of themes for district level resource atlas the guidelines issued by the Planning Commission, Government of India in 1969, 1984 and 1990 have also been considered here.

4.2 EVALUATION OF NATIONAL ATLAS OF INDIA

The National Atlas of India was published in 1982 by National Atlas and Thematic Mapping Organization, Calcutta. The table of contents indicates 300 plates but it contains only 286 plates arranged in eight volumes *viz.*, general and geo-political, physical and geo-morphological, climatic, hydrographical and bio-geographical, population and transport, land utilization and industrial, agricultural, livestock, commercial and economic, social and regional maps and historical besides cultural (Appendix Va). Eight scales for drawing plates have been employed in this atlas. (Table 4.1).

In this atlas, majority of the plates have been drawn in the scales of 1:1 m (33 per cent of the plates), 1:6 m (30.5 per cent), and 1:2 m (28 per cent). Plates in the scale of 1:6 m are single sheet plates. In 1:1 m scale, India is covered in 15 sheets supplying information at inter-district/ district level. In 1:2 m scale, the country is divided into five plates and they provide information at inter-state/state level. Some of them like crop regions, drainage, soil etc., have been represented in two scales (*viz.*, 1:2 m and 1:6 m).

Volumes on climate and bio-geography and population and transport contain 16 per cent of total plates under each of them. Agriculture and economy and Social and regional volumes have 14 and 15 per cent of plates respectively. All other volumes account for less than 40 per cent of plates. The data on climate and bio-geography and population and transport have been represented in about one-third of the total plates contained in the atlas.

With regard to techniques of presentation, conventional methods have been adopted. Thus, Isopleth method has been employed for relief, rainfall, temperature etc., choropleth for density of population, stock density etc., chorochromatic for land use, agricultural regions, urban land use etc., and statistical graphs and diagrams for locational/ area data. Again, where contents of the subject is wide-ranged, like cropping pattern, four-in-one plates have been drawn to display detailed information.

4.3 EVALUATION OF STATE LEVEL ATLASES

In the present study, two state level atlases have been selected for evaluation *viz.*, Basic Resource Atlas of Tamil Nadu and Resource Atlas of Kerala. Planning Atlas of Andhra Pradesh, which is of earlier date, has not been taken for evaluation.

Table 4.1 : National Atlas of India : Distribution of Plates According to Scale and Theme

Sl. No.	Contents of the Volume	Scale								Total
		1:25000	1:100000	1:1 m	1:2 m	1:6 m	1:12 m	1:15m	1:25 m	
(1)	(2)	(3)	(4)	(5)	(6)	(7)	(8)	(9)	(10)	(11)
1.	General & Political	—	—	58	30	8	—	4	—	100 (8)
2.	Physical and Geomorphological	—	—	38	32	27	3	—	—	100 (12)
3.	Climate and Biogeography	—	—	—	52	27	19	—	2	100 (16)
4.	Population and Transport	—	—	55	19	26	—	—	—	100 (16)
5.	Land Use and Industry	—	—	50	8	34	8	—	—	100 (9)
6.	Agriculture and Economy	—	—	—	36	62	2	—	—	100 (14)
7.	Social and Regional	—	—	40	16	42	2	—	—	100 (15)
8.	Historical and Cultural	16	3	43	23	3	12	—	—	100 (10)
	Total in Percentage	1.6	0.3	33	28	30.5	6	0.3	0.3	100

Note : Percentage in the parenthesis in column No. 11 indicates percentage of plates in particular volume to the total number of plates *i.e.,* 300.

4.3.1 Basic Resource Atlas of Tamil Nadu

This atlas was prepared by the Department of Geography, University of Madras in 1982. It contains 142 maps arranged in ten sections. The first section is devoted to the assessment of land resources, followed by water resources in section II, while biotic and mineral resources figure in sections III and IV respectively. Section V analyses the human resources, while section VI shows sources and extent of irrigation. Various facets of agricultural development have been represented in section VII, while industrial aspects have been shown in section VIII. Section IX had been devoted to show the available infrastructural facilities and the last section portrays the levels of development (Appendix Vb).

Section I: Land Resources

A plate on administrative divisions precedes this section. District, *taluk* and block boundaries and their headquarters have been represented by line and point symbols in Plate No.1. Location of Tamil Nadu state is shown in an inset map of India.

Section I has five plates on land resource. The first plate of this section (Plate No. 2) is on relief, showing various physiographic details like plains, plateau, hills, slopes etc. Contours have been used to portray the relief particulars. Plate No. 3 on Geology shows major and minor lineaments, intrusive complexes and geological formations. Chorochromatic technique and also line symbols have been used here. Plate No. 4 depicts the distribution of soil in the State of Tamil Nadu. There are 53 types of soils grouped under nine categories based on their nature. Slope map (Plate No. 5) gives details about degree of slope. Here, isopleth mapping technique has been used for presenting the data. Plate No. 6 on 'Land capability' gives the extent of physical limitations imposed by Nature as well as interference of man with environment. Simple area symbols have been used in this plate.

All these five (land resources) plates and administrative division plate have been drawn in 1:2 m scale. Hence, they provide information upto inter-district/regional level.

Section II: Water Resources

The various phenomena of water resources are presented in a set of seven plates. Of these, three plates have been drawn to show annual rainfall, seasonal rainfall and variability of rainfall. Isopleth technique is used for drawing these maps. The information about areas of high, medium and low rainfall can be identified from the plates. Plate No. 8

on 'Seasonal rainfall' is a four-in-one plate. It shows the contribution of rains by monsoons. A plate on 'Variability of rainfall'(No. 9) has been drawn in view of the impact of high variability of rainfall on agricultural production. A four-in-one plate No.10 has been included in this section in order to reduce the wastage of water through transpiration, to identify the areas of water surplus or deficiency and areas of aridity. Isopleth technique with range graded symbols has been used in this plate.

Plate No. 11 shows the tracks of cyclonic storm and affected areas. The qualitative line symbols represent tracks of storm, while chorochromatic technique shows the categories of the storm affected areas.

Drainage is presented in Plate No. 12. It shows the course of the river, number and pattern of streams etc. Besides showing the network of streams, it also shows location of tanks, with canals and river basins marked distinctly by suitable qualitative point and line symbols. Ground Water plate No.13 is two maps in one plate, wherein one map shows 'Ground water level' and the other 'Ground water quality' in 1:3 m scale. Except this, all others in this section are in 1:2 m scale. Thus, these plates provide water resource information at inter-district level.

Section III: Biotic Resources

This section has six plates giving biotic details. Plate No. 14 shows, distribution of vegetation including natural vegetation (both the types of wet and dry vegetation) and plantations. The data are represented by chorochromatic technique in which different colours and line patterns have been used. Plate No. 15 on 'forests' is another chorochromatic plate showing forest, division-wise kind of plantations including coconut. Both the plates are in 1:2 m scale.

The data on forest revenue received from various forest products are presented in four-maps-in-one plate No. 16. In the first map, data on production of timber, sandal wood and pulp wood are available while in the second map, data on production of fuelwood are represented by bar diagrams. Minor forest produce of different kinds have been represented by proportionate circles and sectors in third map. The fourth map has been drawn to show forest revenue by using divided proportionate squares. In plate No. 15 and 16, the information on forestry is available only at forest divisional level.

Plate No. 17 has four maps (1:4 m scale) showing particulars about milch animal, live stock other than milch animal, dairy projects and poultry units. In the first and second maps, *taluk* level data on milch animal and other than milch animal per 1000 hectares have been represented by choropleth technique, while in third and fourth maps locational details of dairy projects (by proportionate circles and squares) and poultry units (by bar diagrams and proportionate circles) are shown. From the first two maps, information can be generated upto *taluk* level and from the last two maps locational level.

Wild life and sanctuaries are presented in Plate No. 18. It is a 1:2 m plate. The location of different kinds of existing wild animals are shown by their species. Additional information presented along with wild life are available from area under forest and location of sanctuaries. This plate gives information only at locational level.

The last Plate (No. 19) of this section is on 'Marine fisheries'. It is a four-in-one plate drawn in 1:4 m scale. The first map has been drawn to show marine fish production by proportionate circles. In the second map, bar diagrams have been used to portray the available fishing equipments in all coastal districts. Infrastructural facilities like fish processing, fish meal, canning units, boat construction and fish farms have been represented by proportionate squares in four different colours in third and fourth maps. The last map is an enlarged portion of Madras-Chenglepet districts. An inset has been drawn to show varieties of fish, caught in Tamil Nadu state. The first two maps show district-wise details, and the last two maps at locational level.

Section IV: Mineral Resources

There is only one plate on 'Mineral resources', portraying the location of mines and also the type of minerals found in them. Located point symbols have been distinctly marked to show the mines. In the inset, district-wise production of minerals in terms of rupees is represented by proportionate circles. The main map provides location of mines and the inset map gives the district-wise production of minerals in terms of money value.

Section V: Human Resources

This section contains seven plates revealing various aspects of human resources. Plate Nos. 21 and 22 are on distribution of population and vital rates are in 1:2 m scale, while all other plates of this section accommodate four maps in one plate each and are drawn in 1:4 m scale.

In the plate on distribution of population, rural population has been represented by quantitative dots and urban population by proportionate circles. Plate No. 22 shows vital rates of population of Tamil Nadu state. By using multiple line graph, district-wise birth, death and infant mortality rates have been represented in this plate and natural increase in population is shown by colour shading. From these maps, *taluk* and district level information respectively can be generated.

A four-maps-in-one Plate (No. 23) has been drawn to show the growth of population during 1961-71, 1971-81, (*taluk*-wise) and 1951-81. The first two maps and also the last one are drawn using district level data, while the third map is drawn with *taluk* level data. Range graded colour shading was used in all these four maps.

Plate No. 24 has been devoted to represent four different attributes of population *viz.*, density, rural-urban composition, sex-ratio and migration. While the first and third maps have been drawn by choropleth technique, second and fourth maps are drawn with proportionate pie-diagrams. District level data have been used in second, third and fourth maps while in the first map, *taluk* level data are used. Migration map is drawn with 1971 data, and other maps represent the data for 1981. Data on net migration have been represented by flow diagram in four-maps-in-one Plate No. 25 covering Coimbatore, Madras-Chenglepet, Madurai and Tiruchirapalli districts.

Plate No. 26, four-maps-in-one, has been designed to bring out the age-structure, illiterate population, general and also female literacy. All these maps have been drawn with district level data. Age-structure of population has been represented by pyramidal bars, illiterates by multiple bars and general and female literacy by choropleth technique.

Occupational characteristics of workers at district level are shown in four-maps (Plate No. 27). The first map gives an idea of the relative share of different occupational groups in the total population *viz.*, main workers, marginal workers and non-workers and they are represented by pie-diagram. While the other three maps give information on the percentage share of agricultural, industrial and other workers to total main workers, these three maps have been drawn based on *taluk* level data, using choropleth technique. Plate Nos. 26 and 27 are in 1:4 m scale.

Section VI: Irrigation

The sources and extent of irrigation are shown in three plates of this section.

Different sources of irrigation have been shown in plate No. 28. This map gives information on distribution of river, tanks, dams, wells, canals and command areas including those in existence, under construction and also those proposed in future. Plate No. 29 depicts the extent of irrigation showing source-wise net area irrigated by pie-diagram, using *taluk* level data. The inset map portrays the changes in the extent of irrigated area from 1967-68 to 1977-78. It has been drawn by choropleth technique on district level data. These two Plates (Plate Nos. 28 and 29) are in 1:2 m scale. Percentage of source-wise (canal, tank and well) irrigated area to total irrigated area and net irrigated area is presented in a four-maps-in-one Plate (No. 30). Tank, well, canal and irrigated area are presented in first three maps, while percentage of net irrigated area to net cultivated area is found in the fourth map. All these four maps have been drawn applying choropleth technique and are based on *taluk* level data.

Section VII: Agricultural Development

This section contains 17 plates. Plate No. 31 deals with pattern of land use. *Taluk* level land use data have been represented through proportionate pie-diagram. It is in 1:2 m scale. Plate No. 32 is a four-maps-in-one plate, depicting net cultivated area, intensity of cropping, man-land ratio and land holdings of small and marginal farmers. First two maps have been drawn based on *taluk* level data while last two maps are drawn with district level data. All these maps are drawn by choropleth mapping technique.

Plate Nos. 33 to 39 are four maps in one plate and scale of these maps is 1:4 m. Plate Nos. 33, 34, 35 and 36 have been designed to show the area under important crops *viz.* rice, sorghum, *bajra*, pulses, groundnut, other oilseeds, sugarcane and cotton. These maps provide information on the area under these crops at *taluk* level. Relative yield and spread of above mentioned crops have been depicted by choropleth technique in plate Nos. 37 and 38. District level data have been used to draw these maps. Plate Nos. 39 and 40 portray the changes in the area under these crops from 1967-68 to 1977-78. Choropleth technique is applied for mapping *taluk* level data.

Plate No. 41 depicting 'crop regions' has been designed to show crop diversification. While choropleth technique has been used to represent the number of crops, choroschematic method has been used to represent the type of crops. *Taluk* level data have been used for drawing this 1:2 m scale plate.

Plate Nos. 42 and 43 accommodate two maps in each plate. In Plate No. 42, number and area of land holdings have been represented by proportionate pie-diagram. Similarly, Plate No. 43 has two maps on number as well as area of land holdings under tenant cultivation. Divided proportionate squares have been used to represent the data. *Taluk* level data have been used to draw these two plates.

Spatial distribution of periodic markets, regulated markets and cattle shandies have been shown in Plate No. 44, using qualitative point symbols. Different types of agricultural credit societies have been represented in plate No. 45 by proportionate pie-diagram and are based on *taluk* level data. Plate No. 46 has been designed to show distribution of state seed farms. The circles are proportionate to area of seed farms and alphabets represent the type of seed. All these three plates are in the scale of 1:2 m.

Data on agricultural machinery are presented in four maps in one Plate (No. 47). District level data on Bullock and manually operated implements, tractor and tractor operated implements, oil engines, electrical pumpsets and sprayers and sugarcane crushers, oil ghanis and threshers are presented in all these four maps and are depicted by bar diagrams.

Section VIII: Industrial Resources

Five Plates (Nos. 48 to 52) have been devoted to present the data on industrial production and employment.

Plate No. 48 has been designed to portray distribution of power stations and power grid. Proportionate circles have been used to show the type and capacity of power stations, while power grid is represented by qualitative line and point symbols.

Plate No. 49 shows the distribution of industries in four maps. Qualitative dots have been used to differentiate four groups of industries *viz.*, food processing and textile industries, wood, paper and chemical products, metallic and non-metallic products and machinery, miscellaneous industries and repairing industries in 1:4 m scale. Alongwith the industrial data, network of important roads have been added in these maps.

District-wise data on industrial production have been represented by proportionate squares in Plate No. 50. The different types of industries are shown distinctly by patterns. Distribution of industrial employment has been portrayed in Plate No. 51. Proportionate circles represent size of employees. Plate No. 52 shows different types of qualitative point symbols representing the data on tourist centres. All these three plates are drawn in 1:2 m scale.

Section IX: Development Infrastructure

Plate No. 53 has been designed to show roads (national highways, state highways and other major roads), railways (broad gauge and meter gauge) using conventional line symbols and accessibility of roads by isolines. Quantitative line symbols have been used in Plate No. 54 on Bus traffic flow per day. Plate No. 55 is a two in one plate showing the inter-state and also inter-national connection using quantitative line symbols for airways and sea routes, besides showing location of air-ports, major and minor ports represented by qualitative dot symbol.

Plate No. 56 is a four-maps-in-one plate, showing jurisdiction of school education districts, number of school age-group children per school at district level, distribution of centres of higher education with university jurisdiction and distribution of the strength of students in different centres of higher education. Excepting the first map, all other maps provide information at district level. The first map deals with boundaries of school educational districts. In first and third map data have been represented by qualitative line and point symbols and in second map, by choropleth technique and in last map, by proportionate pie-diagram.

The data on post offices, telegraph grid and telegraphic services and trunk call services are presented in Plate No. 57. It is a four-maps-in-one plate, in which first and last maps have been drawn for postal divisions. Post offices and trunk call services have been represented by proportionate pie-diagrams and proportionate squares respectively in second and fourth maps. Map on telegraph grid has been drawn using qualitative point and line symbol. Postal and telegraphic services are represented by horizontally arranged bar diagrams. These maps provide data at postal division level (first and last map), telegraph division level (second map) and state level (third map).

Plate No. 58 shows the distribution of hospitals in the main map and distribution of primary health centres in an inset map. In the main map, circles are proportionate to number of beds and different colours have been used to show different types of hospitals. In the inset, primary health centres are shown by point symbols. Banking services have been represented by proportionate circles and type of offices by different patterns in Plate No. 59. Both these maps give only location-wise information and they are in 1:2 m scale.

Section X: Regional Resource Development

This section has only one plate to portray *i.e.* the levels of regional resource development. It is a four-maps-in-one plate. First three maps give levels of development in agriculture, livestock and industry and the last map provides the regional resource development at the district level.

4.3.2 Resource Atlas of Kerala

This atlas has been brought out by Centre for Earth Science Studies, Trivandrum in 1984. It has 57 plates arranged in five sections, besides a plate on Location and Administrative Divisions of Kerala (Plate No. 1). Section I, deals with natural resources. Various parameters of human resources are presented in section II. Section III is devoted to details of infrastructural facilities. Resource development in different sectors of the economy has been portrayed in section IV while section V gives particulars about regions and also regional development (see Appendix Vc).

Plate No. 1 is a 1:1 m scale plate, with an inset showing the location of Kerala State in 1:24 m scale. This plate has been drawn with qualitative point and also line symbols. *Taluk* level data regarding boundary, headquarters and name of important rivers and reservoirs can be generated from this plate.

Section I: Natural Resources

This section has 14 plates (Nos. 2 to 15), covering natural resources like physiography, drainage, soil, climate etc.

Physiographic details have been portrayed in plate No. 2, which is a 1:1 m plate with an inset in 1:2 m scale. Altitudinal and bathymetric details have been represented by isopleths and bright colour shading and spot height and pass by point symbol. The inset map gives generalised altitudinal zone such as highland, midland and lowland.

Plate No. 3 shows the rivers and canal network, reservoir, gradient of rivers in five different gradients besides river basin boundary. Qualitative line symbols have been predominantly used to represent the above details, while Quantitative line symbols represent river gradients.

Slope particulars have been represented by chorochromatic mapping technique in plate No. 4 in which various degrees of slope and major rivers and reservoir are presented distinctly by colour shading with qualitative description.

Age-wise distribution of geological formations, intrusives and availability of minerals are presented in plate No. 5. Chorochromatic technique and qualitative point symbols have been used for representing the data. In addition to geological details, major roads, railways and district boundaries have also been added in the plate. This map provides information upto district-level only.

In Plate No. 6, hydrogeological data *viz.*, sediment, laterites, dykes, fracture/shear zone and precamprian crystalline and associated intrusives and acquifer area of artesian belt, spring, wells with yield per hour have been presented with the use of chorochromatic, linear and qualitative and quantitative point symbols. This plate has an inset map showing ground water provinces for tapping underground water by various means like heavy duty tube wells, dug wells or domestic wells.

Plate No. 7 portrays the soil types through chorochromatic technique. The legend gives the description of soil types in technical language and also locally prevalent common names. Plate Nos. 2 to 7 are in 1:1 m scale and inset maps in 1:2 m. This section also has 7 plates (Nos. 8 to 14) showing the climatic data. Plate No. 8 portrays the mean monthly temperature using located multiple line graph. Each graph has been designed to show highest, mean-maximum, mean-monthly, mean-minimum and lowest temperature of a particular place distinctly by five different colour lines. Isopleth mapping technique has been used to represent mean annual temperature in the inset map.

Plate No. 9 represents the annual rainfall by isohytes and mean monthly rainfall using located bar diagrams. Isopleth technique has been used to show South-West monsoon (average) rainfall in plate No. 10. The inset map shows the average number of rainy days during this period, represented by located bar diagrams. Average rainfall received during North-East monsoon period has been portrayed in Plate No. 11 using isohytes and average number of rainy days is represented by located bar diagrams. Rainfall received during the period other than these two monsoon seasons (January to May) has been depicted in Plate No. 12 with the help of isohytes. During this period, average number of rainy days has also been represented by located bar diagram in the inset map.

Plate No. 13 is a four-maps-in-one plate, having four 1:2 m maps. First three maps show the variability of rainfall for three seasons — South-West monsoon, North-East monsoon and dry season — while the last map depicts the variability of annual rainfall. In all these four maps, isolines have been used to represent the data.

Data on speed and direction of winds have been represented in plate No. 14 by located wind roses. The bars are proportionate to windy days and they have been segmented to show the wind speed in kilometre per hour. At the centre of the wind rose, percentage of number of calms at 8.30 and 17.30 hr to the total number of days have been written.

Plate No. 15 on vegetation is the last map of this section. This plate depicts natural vegetation types along with adjacent plantations. This data has been represented by chorochromatic technique.

This section provides generalised information about resources at district level and in some cases, their location also.

Section II: Human Resources

Various parameters of human resources have been presented in seven plates (Nos. 16 to 22) of this section.

Data on rural population have been represented in plate No. 16 by quantitative dots. This plate is capable of supplying the information about distribution of rural population upto *taluk* level. Plate No. 17 portrays distribution of urban population along with urban outgrowth by using proportionate spheres. The distribution and location-wise information on size of urban centres can be gathered from these maps. Both these plates are in 1:1 m scale.

In the remaining maps of this section, choropleth technique has been employed representing *taluk* level data. Plate No. 18 is a four-maps-in-one plate showing density, decadal variation in density, sex-ratio and decadal variation in sex-ratio. All these maps have been drawn, applying range graded colours in 1:2 m scale. Percentage of Scheduled Castes to total population has been presented in Plate No. 19. The inset map is on Scheduled Tribes. The particulars of literacy have been presented in Plate No. 20. Themes of this four-maps-in-one plate are literacy, decadal variation of literacy, male literacy and female literacy. Plate No. 21 is devoted to presentation of cultivators, agricultural labourers, household industries and other workers. Data on main workers, female main workers and marginal workers are represented in Plate No. 22.

Section III: Infrastructure

This section has nine Plates (Nos. 23 to 31). Plate No. 23 is on irrigation showing command areas of reservoirs and source-wise irrigated area in an inset map. In the main map, qualitative lines have been used to represent river network and qualitative symbols are used for representing reservoirs and chorochromatic technique is used to show the command areas of the projects already completed, those under execution and also those proposed. Pie-diagram has been employed in the inset map for representing source-wise irrigated area. Region-wise and district-wise information can be generated from the main and inset maps respectively.

Plate No. 24 shows the distribution of power stations, power generated and power grid in the main map. Power station and power generation have been represented by proportionate circles, while transmission lines are represented by qualitative line symbols. There are two bar diagrams giving information about growth of power generation for the past 30 years and also growth of transmission lines from 1967 to 1980 for 220 KV, 110 KV and 66 KV. Data on consumption of electricity in units are presented by proportionate circles.

Transport network (Plate No. 25) represents the network of roads (national highways, state highways and other roads), railways (broad and metre gauge and proposed broad gauge) and air route in qualitative line symbols, while airports and sea ports are noted by qualitative point symbols. There is an inset map giving information about accessibility of roads represented by isopleths. Plate No. 26 is a four-maps-in-one plate, portraying four themes of data *viz.*, inland waterways, railway accessibility, road density and road availability. In the first map, qualitative line symbols have been used to present the inland waterways. The second map uses line symbol for railways and isopleths for accessibility. Last two maps have been drawn by choropleth technique. While first and second maps give region-wise information, third map gives *taluk* level and fourth map district level information respectively.

Distribution of post and telegraph offices is presented in Plate No. 27, using quantitative dots in two colours. In the inset, details of the population served by one post office is presented using choropleth technique on district level data.

Plate No. 28 portrays distribution of banks using quantitative dots/ proportionate circles while inset map shows the number of people served by one bank with the help of choropleth technique. Both main and inset maps supply information about banks at district level. Bank advances to agricultural sector are presented in Plate No. 29. While the number of bank accounts have been represented by proportionate squares, the amount of advance is represented by pie-diagrams. The sectors are proportionate to direct and indirect finances. Different types of cooperative societies, grouped under four categories, are presented in Plate No. 30. Proportionate circles in four different colours have been used to present these categories, while choropleth technique shows the proportion of cooperatives to total cooperatives in the state.

Distribution of schools, colleges, professional colleges and technical institutions are shown in a four-maps-in-one Plate (No. 31), using qualitative point symbols. This plate gives only district level information about different types of centres of education. All Plates of this section in 1:1 m scale and inset maps in 1:2 m, except Plate No. 26 and 31 which are four-in-one plates in the scale of 1:2 m.

Section IV: Resource Development

Sector wise status of resource development is presented in 20 Plates (Nos. 32-51).

Mineral reserves have been represented by proportionate squares and different types of minerals by colours in Plate No. 32. Production of wood in cubic meters are portrayed in Plate No. 33. It is a four-maps-in-one plate. Fuelwood, matchwood, other Indian wood, pulpwood and plywood are the themes of the four maps. All these maps are drawn using choropleth technique on forest division wise data. In the first map, two sets of data are presented by choropleth technique. For one set of data, colours are used and for the other, patterns. This plate gives information on production of wood according to forest division.

A chorochromatic Plate No. 34 shows the general land use. The different types of land use shown in this plate are forest, grass land, arable land, plantation, wasteland, reservoir and lagoon, besides network of rivers and canals. In Plate Nos. 35 and 36, percentage of individual land use types and total area of the district are presented. Forest, permanent pastures and grazing land, land put to non-agricultural uses and barren and uncultivable land are the themes of four-maps-in-one Plate (No. 35). In Plate No. 36, data on land under miscellaneous tree crops, cultivable wasteland, fallow other than current fallow and current fallow are presented. Plate No. 37 is a four-maps-in-one plate, in which percentage of net sown area to total area, area sown more than once, area irrigated and cropping density (total cropped area as per cent of net sown area) are depicted. All these three Plates (Nos. 35, 36 and 37) are drawn using choropleth technique, based on district level data.

Plate Nos. 38 to 44 are showing area and yield particulars of crops like rice, tapioca, coconut, cashew, tea, coffee, pepper, betelnut, sesame, pulses, plantains, sugarcane, banana and rubber. All these maps are of 1:2 m scale. Area under these crops is presented as percentage of total cropped area, while yield per hectare in kilograms. In all these four in one Plates (Nos. 38 to 44), district level data have been represented using choropleth technique. District level data on cropping pattern have been represented by pie-diagram in Plate No. 45. Besides, a crop calendar showing the sowing and harvesting seasons of rice, tapioca, pepper and pulses.

Percentage distribution of livestock has been shown in Plate No. 46, with pie-diagrams. Circles are proportionate to total livestock and the sectors represent cattle, buffalo, goat and others. Percentage value of these categories have been written in the respective sectors. In the inset map dairy units, milk chilling plants, livestock and poultry feed factory, fodder seed farm etc., are presented using qualitative point symbols. Plate No. 47 is a-four-maps-in one plate drawn in 1:2 m scale, based on district level data. First two maps present distribution of duck, drake and duckling. Both these maps use quantitative dots. Third map gives density of poultry population and fourth map availability of poultry per thousand population, by using choropleth technique.

Plate No. 40 on fisheries has four maps in 1:2 m scale. First two maps show fish caught from sea and inland sources respectively, represented by compound bar diagrams and variety of fish shown in different colours. Ice plant, cold storage, canning centres and other infrastructural facilities of fisheries are presented in third map, while fourth map is on fish markets. Both these maps have been drawn by qualitative symbols.

Twelve different types of industries are shown in Plate No. 49 using quantitative, proportionate circles in different colours. It is a 1:1 m plate providing industrial information at *taluk* level. The workers employed in the above twelve types of industries are represented in Plate No. 50. Proportionate circles in various colours have been used to draw this plate. A set of compound bar diagrams has been drawn to show the income of these industries at district level. This plate gives information on industrial workers at *taluk* level and their income at district level. Plate No. 51 is a 1:1 m plate showing the places of tourist importance. The data presented in this plate relate to roads, railways, river network and distance chart, besides tourist centres.

Section V: Region and Regional Development

This section has seven Plates (Nos. 52 to 58) on regions and regional development in 1:1 m scale. First plate of this section shows natural regions like eastern highlands, foothill zones, hilly uplands, rolling plain, palghat gap zone, plains and western low lying plains using chorochromatic technique.

Plate No. 53 is a complex thematic map dealing with rainfall and temperature regions. It also shows pattern of rainfall and potential vegetation of three types classified as moist tropical, tropical attentuated and tropical montane. Chorochromatic technique has been used in this map wherein, climatic data are presented by colour and potential vegetation by using patterns. This Plate gives information at regional level. Crop combination regions are presented in Plate No. 54 by applying choropleth technique. Colours have been used to show the crop regions and alphabets to represent the combination of crops based on district level data. Levels of agricultural and industrial development are presented in Plate Nos. 55 and 56 by choropleth technique respectively. Both these plates provide information only at district level.

Functional hierarchy of urban centres are presented in Plate No. 57. It is a proportionate circle map, showing number of functions. The last plate of this section shows the levels of overall development. Various levels of development like advanced, developed, moderately developed, less developed and backward have been represented by choropleth technique with the use of district level data.

4.3.3 Scale and Level of Data in State Level Atlases

The review of state level atlases of Tamil Nadu and Kerala reveals a variety of scales and levels of data adopted in them. The scales ranged from 1:1 m to 1:24 m and level of data from point location to State. (Table 4.2)

In the state atlas of Kerala, three scales *viz.* 1:1 m, 1:2 m and 1:24 m were adopted, whereas in the atlas of Tamil Nadu, the following six scales — 1:2 m, 1:3 m, 1:4 m, 1:9 m, 1:10 m and 1:12 m were used. In the state atlas of Kerala, the scale of 1:2 m accounts for 70.5 per cent of the total maps drawn, followed by 1:1 m in 28.8 per cent. In the atlas of Tamil Nadu, 1:4 m scale has been adopted in 70.5 per cent of maps, 1:2 m in 19.7 per cent of the maps and the others four scales are used in the remaining maps. Thus, the predominant scale adopted is 1:2 m in the atlas of Kerala and 1:4 m in Tamil Nadu atlas. In the atlas of Tamil Nadu too, the maps printed in 1:4 m scale were originally prepared on a scale of 1:2 m (a scale which is predominant in the atlas of Kerala) and was reduced to 1:4 m for the purpose of printing[1]. Here, it may be pointed out that the size of atlas of Tamil Nadu at 37 cm x 28 cm is more compact than with a size of atlas of Kerala — 45 cm x 38 cm.

With regard to level of data, neither of the two atlases furnished village level data. Atlas of Kerala could reach upto *taluk* level data but it has not covered block level as found in the atlas of Tamil Nadu.

Table 4.2 : State Level Atlases : Distribution of Maps according to Scale and Level of Data

Sl. No.	Scale	Level of Data: Location		Block		Taluk		District		State		Others		Total	
		TN	K	TN	K	TN	K	TN	K	TN	K	TN	K	TN	K
1.	1:1 000 000	—	4.6	—	—	—	4.6	—	19.6	—	—	—	—	—	28.8
2.	1:2 000 000	11.3	0.7	0.7	—	2.1	13.7	4.9	56.1	—	—	0.7	—	19.7	70.5
3.	1:3 000 000	2.8	—	—	—	2.8	—	—	—	—	—	—	—	5.6	—
4.	1:4 000 000	6.3	—	—	—	28.3	—	30.3	—	—	—	5.6	—	70.5	—
5.	1:9 000 000	—	—	—	—	—	—	2.1	—	—	—	—	—	2.1	—
6.	1:10 000 000	—	—	—	—	—	—	—	—	0.7	—	—	—	0.7	—
7.	1:12 000 000	—	—	—	—	—	—	—	—	1.4	—	—	—	1.4	—
8.	1:24 000 000	—	—	—	—	—	—	—	—	—	0.7	—	—	—	0.7
	Total	20.4	5.3	0.7	—	33.2	18.3	37.3	75.7	2.1	0.7	6.3	—	100.00	100.00

* TN - Basic Resource Atlas of Tamil Nadu

** K - Resource Atlas of Kerala

District level information appears to be the predominant level of data in the atlas of Kerala as it is furnished in 75.7 per cent of the maps. In the atlas of Tamil Nadu, district level data account for 37.3 per cent of the maps, *taluk* level data 33.2 per cent and location data 20.4 per cent.

In Tamil Nadu atlas, the predominant scale of 1:4 m provides data largely at district level (30.3 per cent of the maps) and *taluk* level (28.3 per cent). In the atlas of Kerala, the predominant scale of 1:2 m largely furnished data at district level (56.1 per cent of the maps only).

4.4 EVALUATION OF DISTRICT LEVEL ATLAS

In the absence of district level atlases, the only available atlas *i.e.* Planning Atlas of Medak District, has been taken for evaluation in this study.

4.4.1 Planning Atlas of Medak District

The Centre for Economic and Social Studies, Hyderabad published in 1988 a 'Planning Atlas of Medak District', Andhra Pradesh. It is a district level atlas, prepared with the objective of using it not only for preparing the Annual and Five Year Plans of the district but also for adopting a long range strategy for its development[2]. In this atlas, for the first time an attempt was made to use compugraphic output[3]. There are eight sections in this atlas. First section deals with location and administrative division and second section with physical setting. Agricultural and irrigation plates are presented in section III and IV respectively. Section V shows population and section VI socio-economic infrastructure. Maps on rural development programmes and industrial development form sections VII and VIII respectively. There are 106 maps arranged in 57 plates. Of these 18 plates are compugraphic maps (see Appendix Vd).

Section I: Location and Administrative Division

A 1:2 m scale map of India shows the location of Medak district. Plate No. 2 portrays the administrative divisions upto village level alongwith their headquarters. The mandals and villages are numbered and names of *taluk* are written in the respective *taluks*. Both these plates have been drawn using qualitative line and point symbols.

Section II: Physical Setting

Network of streams is presented in Plate No. 3. Perennial streams have been represented by blue lines, non-perennial by black lines while river, sub-basin boundary by brown lines. In the inset, relief details are portrayed with isopleth technique and spot heights. These details are super-imposed with stream network. In this plate, main map is in 1:5 m scale while inset is in 1:1 m scale. This map provides sub-basin level information.

Plate No. 4 is a three in one plate having a main map in 1:5 m and two inset maps in 1:1 m. The main map shows the annual average rainfall by using isohytes and monthly rainfall by located diagram. The two inset maps portray the recharge and run-off. All these maps have been drawn using isopleth technique. *Taluk*-wise information on annual rainfall, monthly rainfall, recharge and run-off can be generated from this plate.

Geology, lineaments and bedrock configuration are the themes of this three-maps-in-one Plate (No. 5). Data on geology and bedrock configuration are presented using chorochromatic technique while lineaments, dykes and quartz veins are shown using qualitative line symbols. Plate No. 6 on hydro-geology portrays the depth of water table with isopleth and water quality by using chorochromatic techniques. Information on depth of water table is available at *taluk* level, while quality of water only at regional level.

Plate No. 7 on soils is four-maps-in-one plate portraying sodium, phosphorous, potassium and chloride in top soil. Classification of soils is presented in Plate No. 8 with an inset on thickness of soil. The maps in these two plates have been drawn by chorochromatic technique. The main map in 1:5 m scale on soil classification provides information at *taluk* level, while other maps in 1:1 m scale at regional level only.

Distribution of land use types is presented in Plate No. 9 using chorochromatic technique. It has been drawn based on satellite imagery. This 1:5 m plate gives information on the distribution of land use at regional level. Plate No. 10 which is a four-in-one plate presents at village level, percentage of area under forest, cultivable waste, net sown area and uncultivable area to the total geographical area by choropleth technique. Plate No. 11 is a map showing index to village numbers for computer generated maps.

Section III: Agriculture

Twelve compugraphic plates (Nos. 12 to 23) are included in this section. All these plates are in 1:5 m scale and they have been drawn with choropleth technique on village level data. First three Plates (Nos. 12 to 14) have been devoted to depict percentage of area under kharif dry crops, rabi crops and principal food crops. Plate Nos. 15 to 19 present data on five crops such as paddy, jowar, bajra, maize and pulses. Area under principal commercial crops has been represented in Plate No. 20. Plate Nos. 21-23 show area under sugarcane, groundnut and castor. These plates give information on agriculture at village level.

Section IV: Irrigation

This section has three plates on irrigation. First plate No. 24 (1:5 m) of this section shows per cent of total irrigated area to total cropped area. The data on source-wise irrigated area *i.e.,* canal, tank, well and other sources are presented in plate No. 25 which is in 1:1 m scale. Both these plates have been drawn using choropleth technique based on village level data. Ground water potential is the theme of plate No. 26. It is a 1:5 m map drawn with chorochromatic technique. Plate Nos. 24 and 25 exhibit information on irrigation at village level and Plate No. 26 regional level respectively.

Section V: Population

Various aspects of population are presented in 12 plates (Nos. 27 to 38) of this section. The data on percentage of decennial variations in growth are depicted in plate No. 27. Plate Nos. 28 and 29 show the density of population for 1971 and 1981 respectively. All these plates are in 1:5 m scale and they provide information at village level.

Plate No. 30 has a main map on the percentage of the Scheduled Castes for 1981 and an inset map on Scheduled Castes population of 1971. Percentage of total Scheduled Tribes population for 1981 is presented in plate No. 31. All these maps are drawn with choropleth technique and they supply information on Scheduled Castes and Tribes at village level.

Data on percentage of literates have been presented in the main map of Plate No. 32. It has an inset map on male literates. This plate has been prepared using choropleth technique.

There are two Plates (Nos. 33 and 34) on occupational structure. The Plate No. 33 has four maps, first two maps are on main workers, of these first map shows total main workers and second map male main workers. Third map is on total marginal workers and the fourth on male marginal workers. Similarly in Plate No. 34, total cultivators and total male cultivators are the themes of first two maps and total agricultural labourers and male agricultural labourers form the theme of the last two maps. Plate No. 35 is two maps-in-one plate. The main map is on total workers in household industry while inset map on male workers in household industry. The data on percentage of total workers employed in other services to total main workers and percentage of non-workers to total population have been represented in two Plates Nos. 36 and 37 respectively. Both these plates are in 1:1 m scale and are drawn with using choropleth technique. All these Plates (Nos. 33 to 37) provide information about occupational structure at village level.

Size-wise distribution of both urban and rural settlement is presented in Plate No. 38. The proportionate circles have been used to represent the size of settlements and colours for differentiating the urban and rural settlements.

Section VI: Economic Infrastructure

This section has four Plates (Nos. 39 to 42) in 1:5 m scale and are devoted to present medical, educational, communication and banking facilities.

Medical facilities which are available in the villages are presented in Plate No. 39. Centres of education from primary school to degree college have been portrayed in Plate No. 40. The data on different kinds of post offices, telegraph offices and telephone offices have been represented in Plate No. 41. Plate No. 42 is on banks such as primary credit societies, primary co-operative agricultural development banks, district co-operative central banks and commercial banks. All these data sets have been represented by qualitative point symbols in different designs and colours.

Section VII: Rural Development Programmes

Facilities created out of rural development programmes are presented in five Plates (Nos. 43 to 47).

First Plate (No. 43) of this section (scale 1:5 m) is on roads and buildings. The main map shows different types of roads and also roads laid by different levels of authorities like zilla parishad, panchayat samithi and local panchayats. All these roads and railways have been represented by qualitative line symbols in different colours. The inset map in 1:4 m scale shows the zilla parishad and panchayati samithi buildings constructed under National Rural Employment Programme by using qualitative point symbols.

Plate No. 44 on a 1:5 m scale plate exhibits protected water supply both under normal scheme as well as mini-scheme and their different stages of implementation by colour shading at village levels. *Taluk*-wise located bars portray the water scarcity-villages, protected water supply, borewell and open wells. Village level information on the protected water supply and also *taluk* level position of drinking water supply can be gathered from this plate.

Houses constructed under rural housing programmes are represented in Plate No. 45. Plate No. 46 is a four-maps-in-one plate and shows houses constructed for Scheduled Castes, Scheduled Tribes, backward classes and economically backward classes. Both these plates have been drawn using proportionate circles in three different colours for differentiating the years. Plate Nos. 45 and 46 provide village-wise information on number of constructed houses, distribution of these houses and also the villages yet to be covered under this scheme.

Plate No. 47 shows number of beneficiaries covered under Integrated Rural Development Programme by choropleth technique and number of beneficiaries and loans advanced to them by located bars.

Plate No. 48 is four-maps-in-one plate. The first map shows the location of grain and livestock markets. Data on the arrival of grain in terms of volume have been shown in the second map using proportionate concentric pie-diagrams where inner circle portrays the quantity and outer circle, the value. Sectors show cereals, pulses, oilseeds, condiments and jaggery. Third map has been devoted to show the arrival of livestock in the livestock markets in terms of volume and value by using two proportionate semi-circles. Estimated market arrivals and the value of grains for seventh plan period are presented in the last map, using proportionate semi-circies. This plate provides information on location of markets and quantity and value of arrival of grain and livestock.

Plate No. 49 is three-maps-in-one plate, showing power grid with sub-stations, by using line and point symbols while two insets are devoted to exhibit the number of agricultural and industrial consumers by using choropleth technique.

Section VIII: Industrial Development

This section has six Plates (Nos. 50-55) devoted to show industrial development. The first plate No. 50 of this section is a three-maps-in-one plate. The distribution, number of workers and capital invested in various industries are portrayed in the main map by proportionate pie-diagram. Distribution of small scale industries is found in one inset and proposed large and medium scale industries in the other.

Plate Nos. 51 to 55 have four maps in each plate. All maps are in 1:1 m scale. Data on number of units and number of workers are presented in one map, using proportionate located circles and choropleth technique respectively. Data on capital invested and value of annual output have been represented in another by proportionate circles and choropleth respectively. These two maps have been drawn for each of the industries covered in this atlas. Each plate portrays two industries and five plates in all cover ten industries.

In Plate No. 51, agro-based and agricultural implements manufacturing industries are presented. Similarly wood based, rubber and leather industries are found in Plate No. 52, chemical industries and ceramics and building materials industries find their place in Plate No. 53, engineering, electrical and allied industries and miscellaneous industries in Plate No. 55. These plates give information on location of industries and also the number of workers employed therein.

4.4.2 Scale and Level of Data in District Level Atlas

The data on distribution of maps based on scale and level of data used have been presented in Table 4.3.

Table 4.3 : Planning Atlas of Medak District : Distribution of Maps According to Scale and Level of Data

Sl. No.	Scale	Level of Data					
		Location	Village	Block	Taluk	State	Total
1.	1: 1 000 000	7.5	30.5	0.9	0.9	—	39.8
2.	1: 5 000 000	7.5	49.0	—	2.8	—	59.3
3.	1:12 000 000	—	—	—	—	0.9	0.9
	Total	15.0	79.5	0.9	3.7	0.9	100.0

In the atlas of Medak district, three scales *viz.*, 1:5 m, 1:1 m and 1:12 m have been used. The village level data account for 79.5 per cent of maps in the atlas. The scale of 1:5 m used for drawing 59.3 per cent of the maps and 1:1 m scale form 39.8 per cent. Most of the maps drawn at these two scales furnish data at village level.

4.5 THEMES OF MAPS BASED ON EVALUATION OF ATLASES

In selecting map themes for district planning, the researcher was guided not only by the direct requirements of district level planning but also factors which may have an indirect bearing on the contents of the district plans. Whereas the possibilities of development in agriculture, animal husbandary and small and village industries etc. will be the contents of district level plans, potentials for development of major activities like mining, large-scale industries, major irrigation projects etc., as found in the district will greatly influence the district level planners in formulating their district level plans and in locating their projects. Thus the themes chosen for district level planning will help and support the planning exercise both directly and indirectly.

The four atlases which have been evaluated, contain 347 themes. They were classified by the researcher into eleven types on the basis of their occurrence in the four atlases as shown in Table 4.4.

At the first instance, the total number of 347 themes have been brought down to 262 through exclusion of 86 themes which have been used only at national level. From the other categories *viz.* 4, 5 and 9, those themes which are relevant to district level planning, have been included in the list. In this way, 118 themes out of 138 themes in these three categories have been included. The themes occurring in the district atlas and atleast in any other (State/National) atlas (categories 1 to 3, 6 to 8) and those figuring in national and state atlases (category 10) were generally relevant to requirements of district level planning and were therefore included in the list (Fig. 4.1).

Table 4.4 : Occurrence of Themes in Atlases

Sl. No.	Categories of Themes by Occurrence	Medak District	Tamil Nadu State	Kerala State	National	No. of Themes
1.	District only	x	—	—	—	34
2.	District and one state	x	x (or)	x	—	26
3.	District and two states	x	x	x	—	7
4.	One of the states	—	x (or)	x	—	92
5.	Two states	—	x	x	—	12
6.	Four atlases	x	x	x	x	18
7.	District and National	x	-	-	x	11
8.	District, any one state and National	x	x (or)	x	x	12
9.	Any one state and National	—	x (or)	x	x	34
10.	Two states and National	—	x	x	x	15
11.	National only	—	—	—	x	86
	Total					347

x occurrence
— Non occurrence

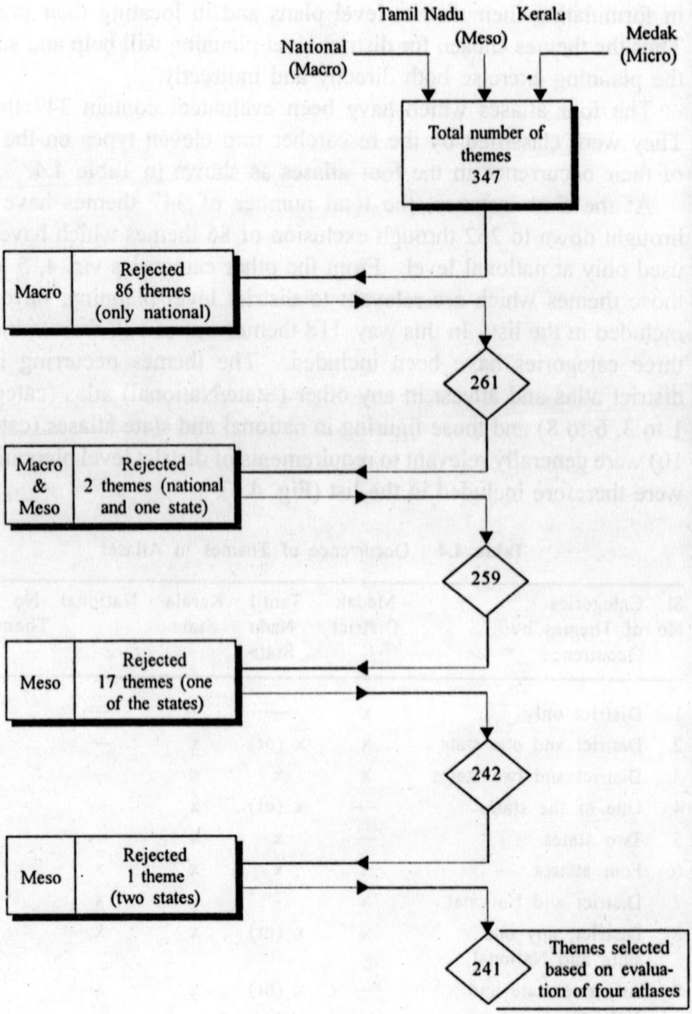

Fig. 4.1 : Themes Selected based on Evaluation of Atlases

Thus, the consolidated list prepared through this procedure consists of 241 themes. Out of the 106 themes excluded, only 86 occur in national atlas. This list of 241 themes can be found in Appendix VI.

4.6 Themes of Map based on Themes of Map based Guidelines of Planning Commission

The Planning Commission of Government of India has issued guidelines for improving formulation and implementation of district plans in 1969, 1984 and 1990. They contain maps on different sectors of economy. The list of maps contained in these three guidelines is presented in Appendix VIIa, b and c.

A review of the list of maps found in the guidelines issued for district planning shows that most of the maps are on location of various facilities and industrial units. Some of these maps also show the distribution of natural phenomena like soil, forest, rivers/streams etc. Maps on population primarily focus on its distribution, composition, concentration of Scheduled Castes/Scheduled Tribes, agricultural labourers etc. A set of maps is meant for the different types of land uses and cropping pattern at block level and also 'problem' areas. Apart from the list of maps suggested in the guidelines of 1969, a list prepared in 1984, contains maps particularly on sources of irrigation, problem areas — such as flooding, water-logging, soil erosion, alkalinity, recurrent drought etc., and also areas having drinking water facilities. There is also a set of maps on projected values of population, agricultural production and other production levels.

Another list circulated in 1990 is a comprehensive one, though some of the maps suggested in 1984 were not included in it *viz.*, problem area, projected values of population, agriculture etc., while new maps on wasteland were added. After a close examination of these lists, the researcher prepared a list of 64 themes of maps for district planning based on the availability of data and also avoided the duplication found in the above said three lists (Appendix VIII).

4.7 Themes and Techniques of the Proposed District Atlas

The following list of 258 themes and techniques has been prepared by the researcher based on the evaluation of atlases at national, state and district levels and perusal of three guidelines issued by the Planning Commission for district level planning in 1969, 1984 and 1990 (Fig. 4.2 and Table 4.5).

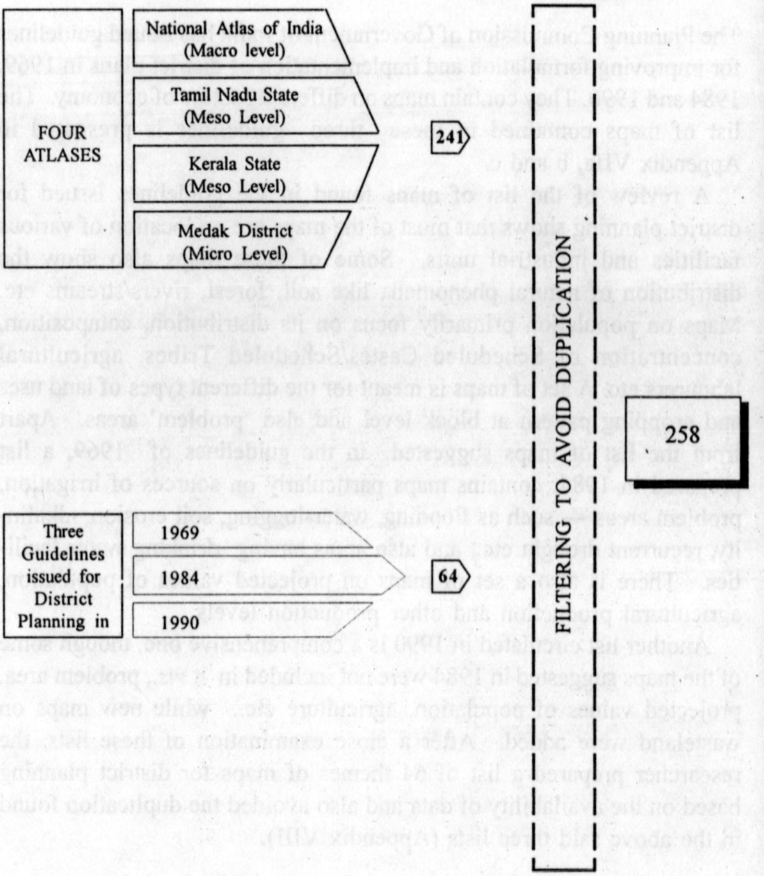

Fig. 4.2 : Themes selected for Circulation among Map users and Designers

Table 4.5 : List of Themes and Techniques for the Proposed District Resourse Atlas based on Evaluation of Atlases and Guidelines of the Planning Commission

Sl. No.	Theme	Technique
1.	Location	Qualitative symbols
2.	Administration	Qualitative symbols
NATURAL RESOURCES		
3.	Relief	Contour
4.	Slope	Isopleth
5.	Drainage	Qualitative line and point symbols
6.	Watershed boundary	Qualitative line symbol
7.	Geology	Chorochromatic technique and qualitative line symbol
8.	Soils	Chorochromatic and choroschematic
9.	Chemical properties of top soil: Sodium	Choropleth
10.	Phosphorous	Choropleth
11.	Potassium	Choropleth
12.	Chloride	Choropleth
13.	Thickness of top soil	Isopleth
14.	Bedrock configuration	Isopleth
15.	Annual rainfall	Isohyte
16.	Monthly rainfall	Isohyte
17.	Seasonal rainfall	Isohyte
18.	Number of rainy days	Bar diagram
19.	Seasonal rainy days	Bar diagram
20.	Mean annual actual evapotranspiration	Isopleth
21.	Mean annual water surplus	Isopleth
22.	Aridity	Isopleth
23.	Mean annual water deficiency	Isopleth
24.	Recharge of water	Isopleth
25.	Water runoff	Isopleth
26.	Variability of rainfall	Isopleth
27.	Annual temperature	Isotherm
28.	Monthly temperature	Isotherm
29.	Wind speed and direction	Wind rose
30.	Natural vegetation	Chorochromatic
31.	Forest	Chorochromatic
32.	Forest revenue	Proportionate square
33.	Forest utilization: Fuelwood	Bar diagram

(Contd.)

Sl. No.	Theme	Technique
34.	Pulp and other Indian wood	Bar diagram
35.	Plywood	Bar diagram
36.	Minor forest produce	Bar diagram
GROUND WATER/IRRIGATION		
37.	Hydro-geology	Chorochromatic
38.	Depth of water table	Isopleth
39.	Ground water quality	Isopleth
40.	Ground water potential	Chorochromatic
41.	Sources of irrigation including location of dams with command area	Chorochromatic, qualitative line and point symbol
42.	Extent of irrigation	Pie-diagram
43.	Irrigated area:	
	Tank	Choropleth
44.	Well	Choropleth
45.	Canal	Choropleth
46.	Other sources	Choropleth
47.	Net irrigated areaChoropleth	
48.	Total irrigated area	Choropleth
49.	Changes in the extent of irrigated area	Choropleth
LAND USE/AGRICULTURE		
50.	Land use	Chorochoromatic
51.	Land capability	Chorochoromatic
52.	Percentage of barren and uncultivable land	Choropleth
53.	Percentage of cultivable waste land	Choropleth
54.	Percentage of fallows other than current fallows	Choropleth
55.	Percentage of current fallows	Choropleth
56.	Percentage of permanent pastures and grazing lands	Choropleth
57.	Percentage of area under non-agricultural uses	Choropleth
58.	Percentage of area under miscellaneous tree crops	Choropleth
59.	Percentage of net sown area	Choropleth
60.	Percentage of area under forest	Choropleth
61.	Distribution of wastelands	Choropleth
62.	Cropping pattern	Pie-diagram
63.	Area under kharif dry crops	Pie-diagram

(Contd.)

Sl. No.	Theme	Technique
64.	Area under rabi crops	Pie-diagram
65.	Area and yield of principal food crops	Choropleth
66.	Area and yield of principal commercial crops	Choropleth
67.	Area under Rice	Choropleth
68.	Area under Bajra	Choropleth
69.	Area under Jowar	Choropleth
70.	Area under Maize	Choropleth
71.	Area under Sorghum	Choropleth
72.	Area under Pulses	Choropleth
73.	Area under Banana	Choropleth
74.	Area under Other plantain	Choropleth
75.	Area under Tapioca	Choropleth
76.	Area under Groundnut	Choropleth
77.	Area under Sesame	Choropleth
78.	Area under Castor	Choropleth
79.	Area under Coconut	Choropleth
80.	Area under Other oilseeds	Choropleth
81.	Area under Sugarcane	Choropleth
82.	Area under Cotton	Choropleth
83.	Area under Tea	Choropleth
84.	Area under Coffee	Choropleth
85.	Area under Pepper	Choropleth
86.	Area under Betelnut	Choropleth
87.	Area under Rubber	Choropleth
88.	Yield of Rice	Choropleth
89.	Yield of Bajra	Choropleth
90.	Yield of Jowar	Choropleth
91.	Yield of Maize	Choropleth
92.	Yield of Sorghum	Choropleth
93.	Yield of Pulses	Choropleth
94.	Yield of Banana	Choropleth
95.	Yield of Other plantain	Choropleth
96.	Yield of Tapioca	Choropleth
97.	Yield of Groundnut	Choropleth
98.	Yield of Sesame	Choropleth
99.	Yield of Castor	Choropleth
100.	Yield of Coconut	Choropleth
101.	Yield of Other oilseeds	Choropleth
102.	Yield of Sugarcane	Choropleth
103.	Yield of Cotton	Choropleth
104.	Yield of Tea	Choropleth
105.	Yield of Coffee	Choropleth
106.	Yield of Pepper	Choropleth
107.	Yield of Betelnut	Choropleth
108.	Yield of Rubber	Choropleth

(Contd.)

Sl. No.	Theme	Technique
109.	Changes in the area under Rice	Choropleth
110.	Changes in the area under Sorghum	Choropleth
111.	Changes in the area under Bajra	Choropleth
112.	Changes in the area under Pulses	Choropleth
113.	Changes in the area under Sugarcane	Choropleth
114.	Changes in the area under Cotton	Choropleth
115.	Changes in the area under Groundnut	Choropleth
116.	Changes in the area under Other oilseeds	Choropleth
117.	Area sown more than once	Choropleth
118.	Intensity of cropping	Choropleth
119.	Man-land ratio	Choropleth
120.	State seed farms	Qualitative point symbol
121.	Storage godowns	Point symbol
122.	Seed and fertilizer shops	Proportionate point symbol
123.	Markets	Qualitative point symbol
124.	Agricultural credit societies	Qualitative point symbol
125.	Arrival of grains to markets (Quantity & Value)	Circles
126.	Grains : Estimated arrival and value	Circles
127.	Crop regions	Chorochromatic and Choroschematic technique
128.	Land holdings of small and marginal farmers	Pie-diagram
129.	Land holding : Area	Pie-diagram
130.	Number	Pie-diagram
131.	Land holdings of tenant cultivators : Area	Pie-diagram
132.	Number	Pie-diagram
133.	Tractors and tractor operated implements	Choropleth
134.	Bullock and manually operated implements	Choropleth
135.	Oil engines, electrical pumpsets and sprayers	Choropleth
136.	Sugarcane crushers, oil ghannies and thrushers	Choropleth

(Contd.)

Sl. No.	Theme	Technique
ANIMAL HUSBANDRY		
137.	Percentage distribution of livestock	Pie-diagram
138.	Milch animal	Choropleth
139.	Dairy projects	Circles
140.	Milk routes	Qualitative line symbol
141.	Poultry farms	Qualitative point symbol
142.	Distribution of poultry population	Qualitative point symbol
143.	Density of poultry population	Choropleth
144.	Availability of poultry/1000 persons	Choropleth
145.	Marine fisheries	Pie-diagram
146.	Inland fisheries	Pie-diagram
147.	Fishing equipments	Bar-diagram
148.	Fisheries infrastructure	Qualitative point symbol
149.	Markets for fish	Qualitative point symbol
150.	Markets for livestock	Qualitative point symbol
151.	Arrival of livestock to markets (Quantity & Value)	Circles
HUMAN RESOURCES		
152.	Growth of population 1971-81	Located graph
153.	Growth of population 1961-71	Located graph
154.	Growth of population 1951-61	Located graph
155.	Density of population 1981	Choropleth
156.	Density of population 1971	Choropleth
157.	Distribution of population	Quantitative point symbol and circle
158.	Migration	Pie-diagram
159.	Total Scheduled Castes Population 1981	Choropleth
160	Total Scheduled Castes Population 1971	Choropleth
161.	Total Scheduled Tribes population	Choropleth
162.	Sex ratio	Choropleth
163.	Decadal variation of sex-ratio	Choropleth
164.	Literates	Choropleth
165.	Male literates	Choropleth
166.	Decadal variation in literacy	Choropleth
167.	Birth, death and infant mortality rates	Located graph

(Contd.)

Sl. No.	Theme	Technique
168.	Rural-Urban composition	Pie-diagram
169.	Age-structure of population	Pyramidal diagram
170.	Occupational structure	Pie-diagram
171.	Percentage of Workers	Choropleth
172.	Percentage of Non-workers	Choropleth
173.	Percentage of cultivators	Choropleth
174.	Percentage of male cultivators	Choropleth
175.	Percentage of agricultural labourers	Choropleth
176.	Percentage of male agricultural labourers	Choropleth
177.	Percentage of household industries	Choropleth
178.	Percentage of male workers in household industry	Choropleth
179.	Percentage of marginal workers	Choropleth
180.	Percentage of male marginal workers	Choropleth
181.	Percentage of main workers	Choropleth
182.	Percentage of male main workers	Choropleth
183.	Percentage of other workers	Choropleth
184.	Distribution of urban and rural settlements	Circle
185.	Functional hierarchy of urban centres	Circle
RURAL DEVELOPMENT		
186.	Construction of buildings under National Rural Employment programme	Qualitative point symbol
187.	Supply of drinking water : Sources and quality of water	Qualitative point symbol and Pie-diagram
188.	Drinking water problem villages	Qualitative point symbol
189.	Rural works of Panchayat Union	Qualitative point symbol
190.	Housing for Scheduled Castes	Circle
191.	Housing for Scheduled Tribes	Circle
192.	Housing for Backward Classes	Circle
193	Housing for Economically Backward Classes	Circle
194	IRDP loans advanced to beneficiaries under various schemes	Bar diagram

(Contd.)

Sl. No.	Theme	Technique
INDUSTRIES		
195.	Power grid with transformers	Qualitative line and point symbol
196.	Villages to be electrified	Qualitative point symbol
197.	Power generation	Circle
198.	Power consumption	Circle
199.	Location of mines	Qualitative point symbol
200.	Production of minerals	Circle
201.	Large and medium scale industries: Workers employed and capital invested	Pie-diagram
202.	Distribution of small scale industries	Qualitative point symbol
203.	Agro-based Industries: No. of units and workers	Circle
204.	Capital invested and value of output	Circle
205.	Agri. implements manufacturing industries: No. of units and workers	Circle
206.	Capital invested and value of output	Circle
207.	Wood-based Industries: No. of units and workers	Circle
208.	Capital invested and value of output	Circle
209.	Rubber & Leather Industries: No. of units and workers	Circle
210.	Capital invested and value of output	Circle
211.	Chemical Industries: No. of units and workers	Circle
212.	Capital invested and value of output	Circle
213.	Ceramic & Building Materials Industries No. of units and workers	Circle
214.	Capital invested and value of output	Circle
215.	Engineering Industries: No. of units and workers	Circle
216.	Capital invested and value of output	Circle
217.	Electrical & Allied Industries: No. of units and workers	Circle

(Contd.)

Sl. No.	Theme	Technique
218.	Capital invested and value of output	Circle
219.	Printing, binding and allied Industries: No. of units and workers	Circle
220.	Capital invested and value of output	Circle
221.	Miscellaneous Industries: No. of units and workers	Circle
222.	Capital invested and value of output	Circle
223.	Location of industrial estates	Qualitative point symbol
224.	Proposed large and medium scale industries	Qualitative point symbol
225.	Tourism	Qualitative point and line symbols
226.	Wild life and sanctuaries	Qualitative point and area symbols
227.	Places of religious and historical importance	Qualitative point and line symbols
INFRASTRUCTURE FACILITIES		
228.	Roads and railways	Qualitative line symbol
229.	Inland waterways	Qualitative line symbol
230.	Airways and sea routes	Quantitative line symbol
231.	Bus traffic flow	Quantitative line symbol
232.	Centres of education	Qualitative point symbol
233.	School education districts	Qualitative line symbol
234.	Students strength of centres of education	Circle
235.	Banks	Qualitative point symbol
236.	Service area of the banks	Qualitative area symbol
237.	Population served per bank	Choropleth
238.	Telegraph grid	Qualitative point and line symbol
239.	Road accessibility	Isopleth
240.	Road availability	Choropleth
241.	Road density	Choropleth
242.	Railway accessibility	Isopleth
243.	Post and telegraph offices	Qualitative point symbol
244.	Population served by post office	Choropleth
245.	Hospital	Qualitative point symbol
246.	Primary health centres	Qualitative point symbol
247.	Veterinary facilities	Qualitative point symbol

(Contd

PROBLEM AREA

48.	Problem areas — Areas liable to flooding	Chorochromatic
49.	Water-logging	Chorochromatic
50.	Soil erosion	Chorochromatic
51.	Alkalinity	Chorochromatic
52.	Recurrent drought	Chorochromatic

REGIONS AND DEVELOPMENT

53.	Natural Regions	Chorochromatic
54.	Bio-climatic Regions	Chorochromatic
55.	Levels of agricultural development	Choropleth
56.	Levels of industrial development	Choropleth
57.	Levels of livestock development	Choropleth
58.	Levels of development	Choropleth

In this chapter, after an evaluation of four atlases at national, state and district level the researcher has prepared a list of themes for a resource atlas at district level. Further, the guidelines issued by the Planning Commission in this regard in 1969,1984 and 1990 were also examined and on the basis of this exercise, a comprehensive list of themes with appropriate cartographic techniques has been suggested. This list of themes and also the techniques used are further scrutinised in the next chapter, based on the perception of both the users and the designers.

REFERENCES

1. A. Ramesh and P.S. Tiwari., *Basic Resource Atlas of Tamil Nadu,* Madras University, Madras, 1982, p. 2.
2. Afzal Mohamed, *Planning Atlas of Medak District, Andhra Pradesh, India,* The Centre for Economic and Social Studies, Hyderabad, 1988, p. i.
3. Idem.

5

USER'S AND DESIGNER'S PERCEPTION IN PLANNING AND PREPARATION OF DISTRICT RESOURCE ATLAS

5.1 INTRODUCTION

The present chapter analyses perception of both map users and designers.

The data on use of maps, scale of maps used, and also various requirements of a resource atlas *viz.*, periodicity of updating, level of data to be used, new maps required by them etc., have been collected from 40 officials working at district level offices. The map designers gave their perception about content, scale, level of data, periodicity etc. of resource atlas for the purpose of district planning. These two sets of data collected from the respondents have been analysed in this chapter.

5.2 PERCEPTION OF MAP USERS

5.2.1 Jurisdiction

The data on area of operation of respondents shows that, 95.0 per cent of the respondents' functions cover the entire district and 2.2 per cent of respondents *i.e.* Agricultural Engineering (Western Ghats Development Programme) cover the area of Western Ghats in three districts *viz.*, Coimbatore, Dindigul and Madurai. While the remaining 2.3 per cent of them (Forest Department) cover the district excluding Kodaikanal Forest Division.

It is evident from the above analysis that the district is the area of operation of the majority of the respondents' (95 per cent).

5.2.2 Functions

The major functions of most of the departments working at district level include planning, implementation and monitoring, and other activities (which includes maintenance of building, ground water investigation, enforcement of law and order etc.,). The two important functions are planning and implementation and monitoring, which account for 84 and 89 per cent of the respondents respectively. Further, 35 per cent undertake other functions.

It may, therefore, be concluded that the major functions of more than half of the respondents are project implementation and monitoring, while two-fifths of them do the planning work.

5.2.3 Map Users

The respondents fall in two groups *viz.*, actual map users and potential map users. Of the total 40 respondents, only 41 per cent of them use maps. Welfare departments like Adi-Dravida Welfare, Panchayat Development and District Revenue Office are among those (58 per cent) who do not use maps but whose work need maps and they are therefore, potential map users.

The reasons given by the respondents for not using the maps in their departmental activities are : 1. Financial constraint in buying maps, and 2. Non-availability of required maps. In spite of all these limitations all of them are willing to use maps.

Most of the map users suggested that, they may be provided with a capsule course/orientation programme to impress upon them the need for using maps at district level so as to increase their working efficiency for chalking out better planning strategies at the given level.

(a) Function-wise map users: Data on function-wise distribution of map users can be seen in Table 5.1.

Regarding the major functions and the use of maps. 67 and 47 per cent of respondents use maps in plan implementation and monitoring and planning respectively. Those who use maps for other functions form 20 per cent of the respondents.

Thus the above analysis shows that, project implementation and monitoring and planning are the two major functions which require maps.

(b) Scale of maps: Particulars of the scale of the maps used by the map-using-respondents are presented in Table 5.2. It is clear from the table that 33 per cent of them use 1:63360 scale, while 21 and 27 per cent use 1:250000 and 1:500000 scale respectively. The project implementing departments like Agricultural Engineering (WGDP) and Public Works Department use 1:50 or 100 scale for drawing maps. Only 13 per cent

Table 5.1 : Map Users

Sl. No.	User Department	Plann-ing	Implemen-tation & Monitoring	Other Activ-ties
1.	Agri. Engg. (Soil conservation)	x	x	—
2.	Agri Engg. (WGDP)	x	x	—
3.	Public Works Department	x	x	—
4.	Highways & Rural Works	—	x	—
5.	Ground Water Department	—	—	x
6.	TWAD (RWS)	x	x	—
7.	Public Health	x	—	—
8.	Forest	x	x	—
9.	District Technical Cell	—	x	—
10.	Agriculture	—	x	—
11.	Agri. (Oilseeds)	—	x	—
12.	Agri. (Training and Visit)	—	x	—
13.	Family Welfare	x	x	—
14.	T.N.E.B.	—	—	x
15.	Police	—	—	x
	Total in Percentage	47	67	20

x Users, — Non Users

of them (Agri. Engineering - Soil conservation and Forest) use other scales like 1:16000 and 1:5000 the remaining 7 per cent use no scale because they draw sketches (Department of Public Health).

Therefore, it is evident from the above analysis that about one-third of the respondents use 1:63360 and nearly each of one-fourth of respondents use 1:250000 and 1:500000 scales.

(c) Sources of maps : Particulars relating to the sources from which users derived their maps are furnished in Table 5.3.

Table 5.2 : Scale of Maps Used by the Map Users

Sl. No.	Name of the Dept.	Scale of the Maps						
		1:50/100	1:63360	1:125000	1:250000	1:500000	Others	No. Scale
1.	Agri. Engg. (Soil conservation)	—	x	—	—	—	x	—
2.	Agri. Engg.(WGDP)	x	—	—	—	—	—	—
3.	Public Works Department	x	—	—	—	—	—	—
4.	Highways and Rural Works	—	—	x	—	—	—	—
5.	Ground Water Dept	—	x	—	x	—	—	—
6.	TWAD (RWS)	—	—	—	x	—	—	—
7.	Public Health	—	—	—	—	—	—	x
8.	Forest	—	x	—	—	—	x	—
9.	District Technical Cell	—	x	—	—	—	—	—
10.	Agriculture	—	—	—	—	x	—	—
11.	Agri. (Oilseeds)	—	—	—	—	x	—	—
12.	Agri. (Training and Visit)	—	—	—	—	x	—	—
13.	Family Welfare	—	—	—	—	x	—	—
14.	TNEB	—	x	—	—	—	—	—
15.	Police	—	—	—	x	—	—	—
	Total in Percentage	13	34	7	21	27	13	7

x — Users
— Non Users

Table 5.3 : Sources of Maps

Sl. No.	Sources	Respondents in Percentage
1.	Prepared by user department	100.0
2.	Borrowed from other departments	33.3
3.	Purchased from Dept. of Survey and Land Records and Survey of India	33.3

It is clear from the table that, all the respondents are in a position to prepare maps in their departments. In addition to that, some of them (33 per cent) borrow maps from other offices and (33 per cent) purchase the required maps from Dept. of Survey and Land Records and Survey of India.

(d) Theme of maps used by the map users: Theme of maps which are used by the map-using-departments are presented in Table 5.4.

Table 5.4 : Theme of Maps used by the User Department

Sl. No.	Theme of Maps	User Department
1.	Contour	Agri. Engineering (Soil Conservation)
2.	Location of district	TWAD (RWS), Family Welfare
3.	Watershed and river basins	Agri. Engineering (Soil Conservation)
4.	Soil conservation	Agri. Engineering (Soil Conservation)
5.	Field measurement maps	Agri. Engineering (WGDP)
6.	Location of sand and lime stone quarries	Public Works Department
7.	Location of reservoirs	Public Works Department
8.	Building plans	Public Works Department
9.	Jurisdiction map	Highways and Rural Works
10.	Hamlet-wise population	Highways and Rural Works
11.	Town/*Panchayat* roads	Highways and Rural Works
12.	Ground water level (Block-wise)	District Technical Cell TWAD (RWS) Ground Water Department
13.	Ground water potential	TWAD (RWS) Ground Water Department
14.	Ground water quality	Ground Water Department
15.	Location of health facilities	Public Health
16.	Roads	Public Health

(Contd.)

Table 5.4 *(Contd.)*

Sl. No.	Theme of Maps	User Department
17.	Flood-prone areas	Public Health
18.	Prevalance of diseases in the last five years	Public Health
19.	Forest area	Forest
20.	Encroached area	Forest
21.	Plantation area (including aerial seeding)	Forest
22.	Fire areas	Forest
23.	Jurisdiction of protected area	Forest
24.	Forest settlement (4 notification)	Forest
25.	Forest settlement (16 notification)	Forest
26.	Forest range management map	Forest
27.	Hydro-geology	District Technical Cell
28.	Fertility status of soil	Agriculture
29.	Area coverage (oilseeds)	Agriculture (Oilseeds)
30.	Rainfall	Agriculture (Oilseeds) Agriculture (Training & Visit)
31.	Cropping pattern	Agriculture (Oilseeds)
32.	Location map (showing office of Agri. Depts. and head quarters of Asst. Agri. Officers)	Agriculture (Training & Visit)
33.	Area coverage	Agriculture (Training & Visit)
34.	Distribution of fertilizer, seed and pesticide year-wise	Agriculture (Training & Visit)
35.	Irrigated and unirrigated area	Agriculture (Training & Visit)
36.	Infrastructural facilities	Agriculture (Training & Visit)
37.	Action project map	Agriculture (Training & Visit)
38.	Special project map	Agriculture (Training & Visit)
39.	Couple protection rate	Family Welfare
40.	Birth, death and infant mortality rates	Family Welfare
41.	Power grid	Tamil Nadu State Electricity Board
42.	Police jurisdiction (showing sub-division/circles)	Police

All map using departments have location map showing the location particulars of their department. Ground water level, quality and potential maps are used by TWAD (RWS), Ground Water Department and District Technical Cell. Contour maps are used by Agricultural Engineering (Soil conservation), Agricultural Engineering (WGDP) and Public Works Department. Maps on watershed and river basin are referred by Agricultural Engineering (Soil conservation). Building plans and also location of sand and lime stone quarries are used by Public Works Department for the maintenance of these buildings. For laying new link roads, town panchayat roads, Highways and Rural Works Department use a map on hamlet-wise distribution of population. The theme of maps used by the Public Health Department are flood-prone and diseases-prevalent areas.

The Forest Department is one of the most important map users. This department uses maps on forest area, encroached area, plantation area, forest settlement, protected forest and forest range map. Maps on area under different crops are used by the Departments of Statistics, Agriculture (oilseeds) and Agriculture (Training and Visit). Rainfall maps are being utilised by the Agriculture (oilseeds) and Agriculture (Training and Visit). Besides these, maps on distribution of seeds, fertilizer, pesticide, irrigated and unirrigated area are prepared and used by the Agriculture (Training and Visit) Department. Maps on vital statistics and couple protection rate are used by the Department of Family Welfare.

It is clear from the above analysis that, those departments whose activities take into consideration the spatial aspects use maps.

5.2.4. Perception of Map Users

Interaction by the researcher with these officials about technical aspects of the map like mapping, scale, level of data etc., brought into focus their perception of the maps needed for district planning. The process of ascertaining their perception was facilitated by the perusal of a few maps presented by the researcher to these officers.

(a) Scale : Most of the officials found the scale of 1:250000 and appropriate one for district level maps. A higher scale, say 1:50000 would make the map unwieldly and difficult for use, while a lower scale — 1:500000 – would render maps too small to facilitate plotting of additional information.

(b) Level of data : For understanding the perception of users about the level of data, the researcher presented four maps to these officials – one set of two maps containing revenue village level data and another set of two maps having block level data on the population of Dindigul District (Maps 5.1 to 5.4). Most of these officers could not read the first set of maps containing village level data as they found village level presentation in maps visually incomprehensible. On the other hand, they had no difficulty in grasping the second set of maps presenting data at the block level. Naturally, most of these officers opted for block level data. Moreover, these officers with district level jurisdiction felt that there was no need to go below block level for presentation of data in the maps. Thus, these officers can be considered to be in favour of block level data for the maps.

(c) Technique of mapping : Four maps employing different techniques *viz.*, 1. Choropleth-patterns (Map 5.5), 2. Isopleth (Map 5.6), 3. Chorochromatic-pattern (Map 5.7), and, 4. Diagram-pattern (Map 5.8) were presented to the map users. Thus, a set of maps — black and white maps with pattern was presented to them. Diagram was favoured by almost all officials because of their experience in the use of statistical diagram. Most of them preferred a four-fold classification of data to be presented in the choropleth maps.

(d) Manual/Computer maps : The map users were invited to examine manually-drawn maps and computer-drawn maps. None of them found any difference between the two and could not choose one over the other (Map 5.9 to 5.11).

5.2.5 Case Studies of Selected Map Users

A humanistic attempt has been made here by the researcher, to bring out the actual perception of the selected map users regarding technique, scale, level of data etc. Of the 40 total respondents, only 15 are currently using maps in their routine functions, while the remaining 25 are potential map users for better decision making. Hence, a random sample of one respondent from each group has been selected and their perceptions have been recorded in the following paragraphs.

(a) Respondent 'A' from actual map user : Respondent 'A' is one of the district level officials in charge of agricultural development. His

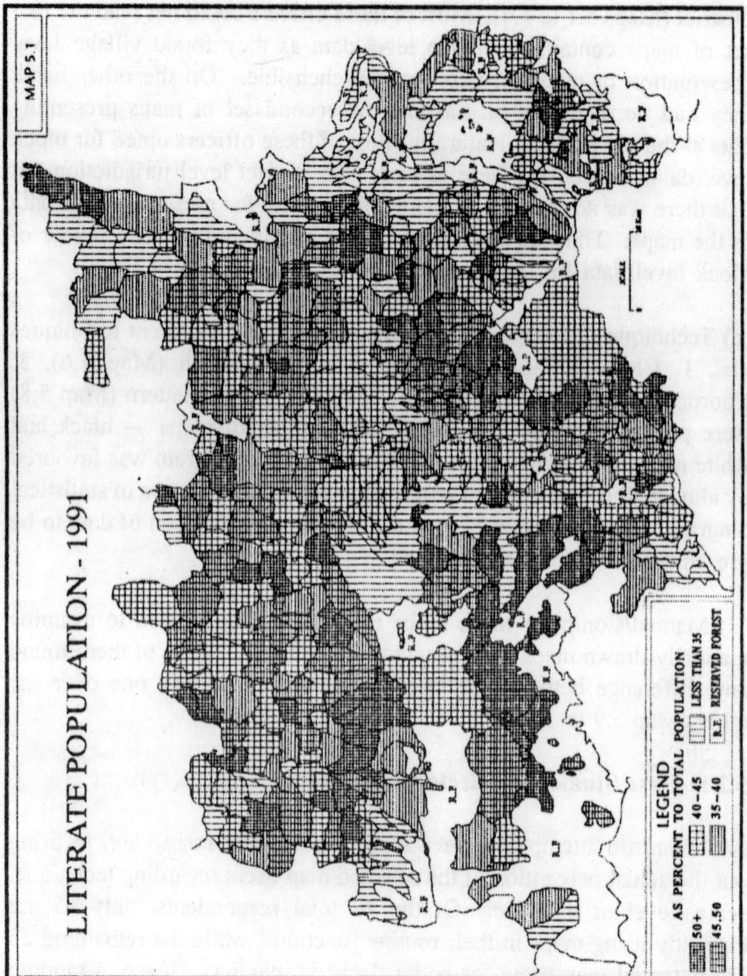
MAP 5.1
LITERATE POPULATION - 1991
LEGEND
AS PERCENT TO TOTAL POPULATION
50+
45.50
40–45
35–40
LESS THAN 35
R.F. RESERVED FOREST

Map 5.1

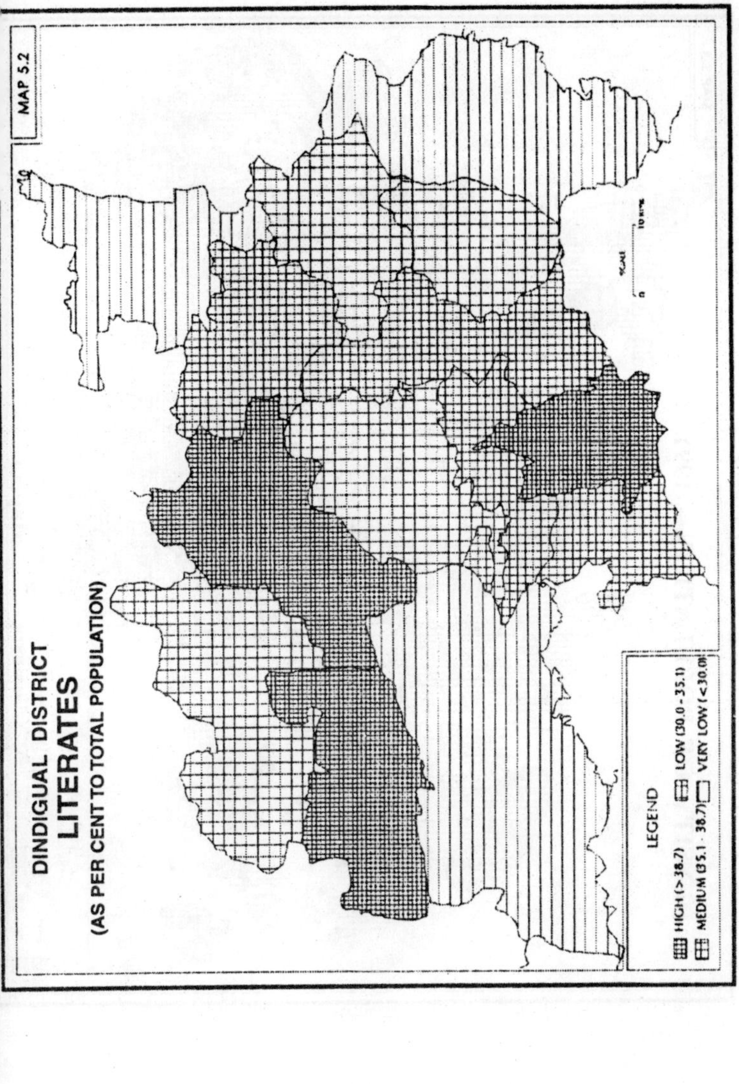
MAP 5.2
DINDIGUAL DISTRICT
LITERATES
(AS PER CENT TO TOTAL POPULATION)
LEGEND
HIGH (> 38.7)
LOW (30.0 - 35.1)
MEDIUM (35.1 - 38.7)
VERY LOW (< 30.0)

Map 5.2

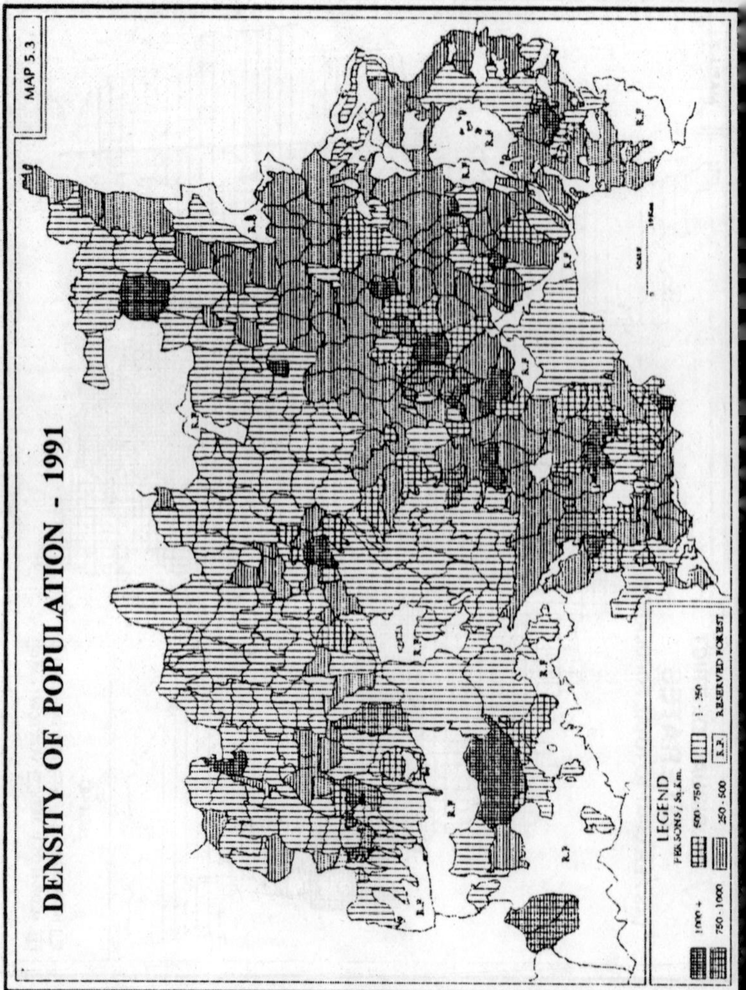
DENSITY OF POPULATION - 1991
MAP 5.3
LEGEND
PERSONS / Sq.Km.
1000 +
750 - 1000
500 - 750
250 - 500
1 - 250
R.F. RESERVED FOREST
R.F

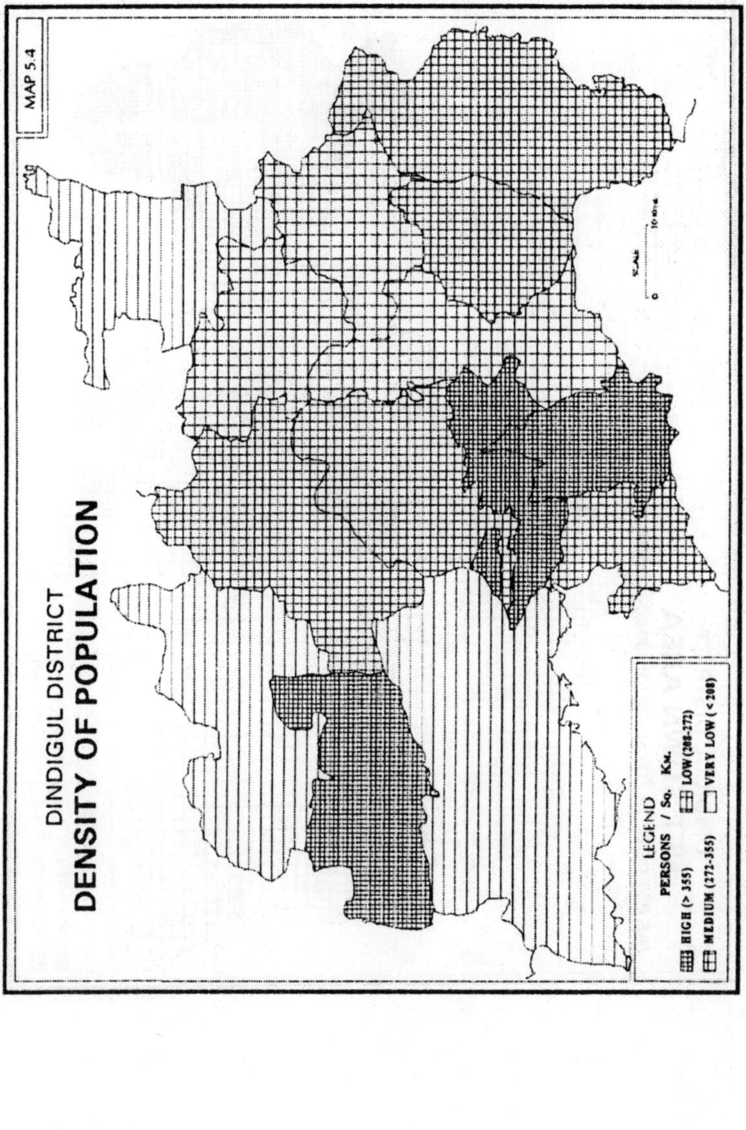
MAP 5.4
DINDIGUL DISTRICT
DENSITY OF POPULATION
LEGEND
PERSONS / Sq. Km.
HIGH (> 355)
MEDIUM (272-355)
LOW (208-272)
VERY LOW (< 208)

Map 5.4

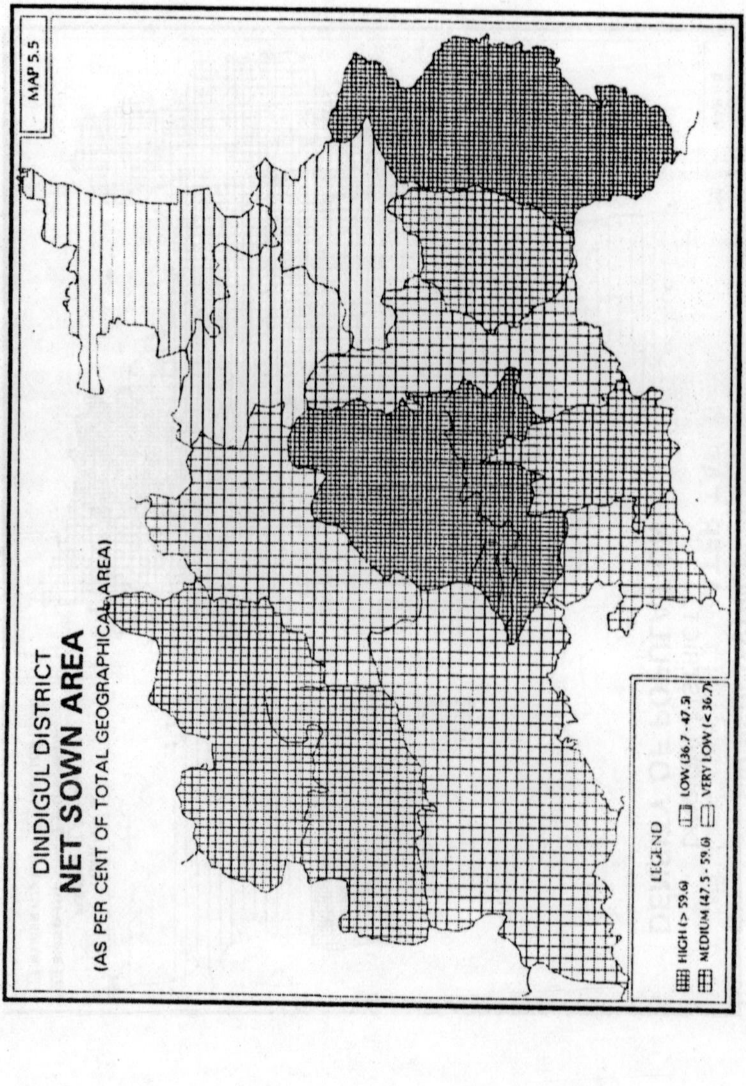
MAP 5.5
DINDIGUL DISTRICT
NET SOWN AREA
(AS PER CENT OF TOTAL GEOGRAPHICAL AREA)
LEGEND
HIGH (> 59.6)
MEDIUM (47.5 - 59.6)
LOW (36.7 - 47.5)
VERY LOW (< 36.7)

Map 5.5

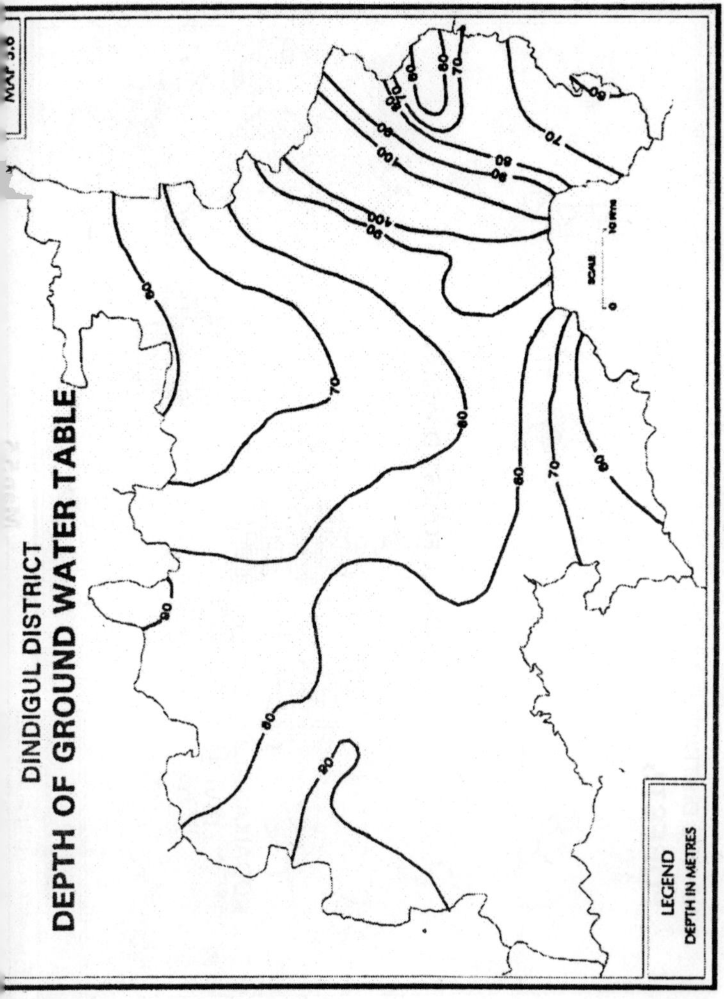
DINDIGUL DISTRICT
DEPTH OF GROUND WATER TABLE
LEGEND
DEPTH IN METRES

Map 5.6

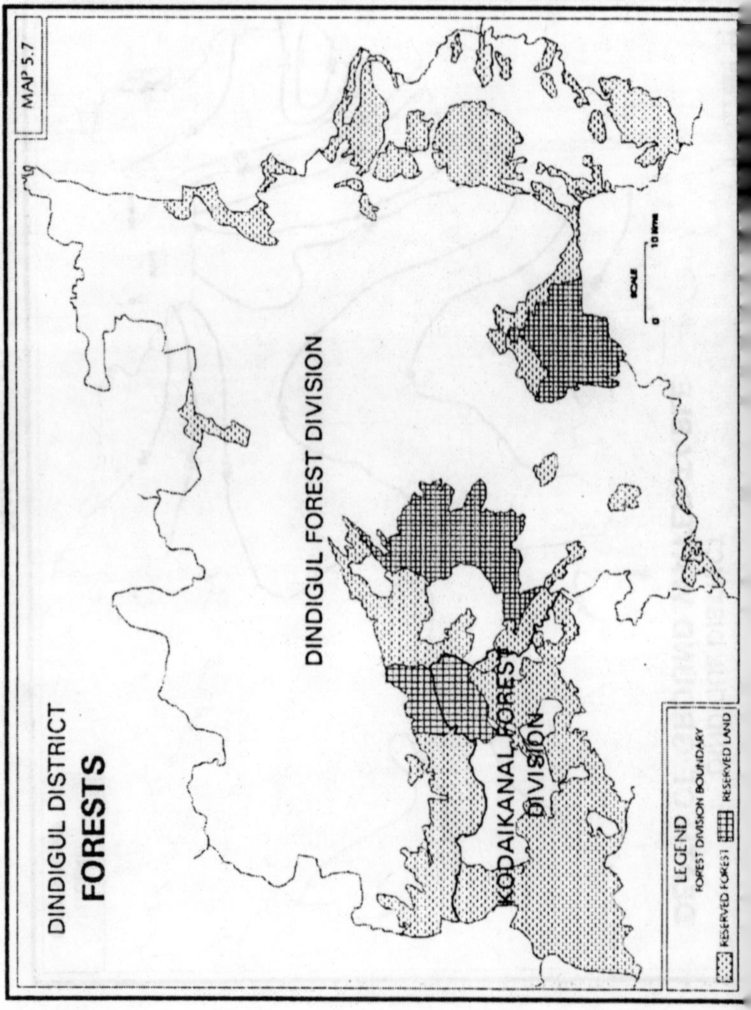
MAP 5.7
DINDIGUL DISTRICT
FORESTS
DINDIGUL FOREST DIVISION
KODAIKANAL FOREST DIVISION
LEGEND
FOREST DIVISION BOUNDARY
RESERVED FOREST
RESERVED LAND
SCALE
0
10 kms

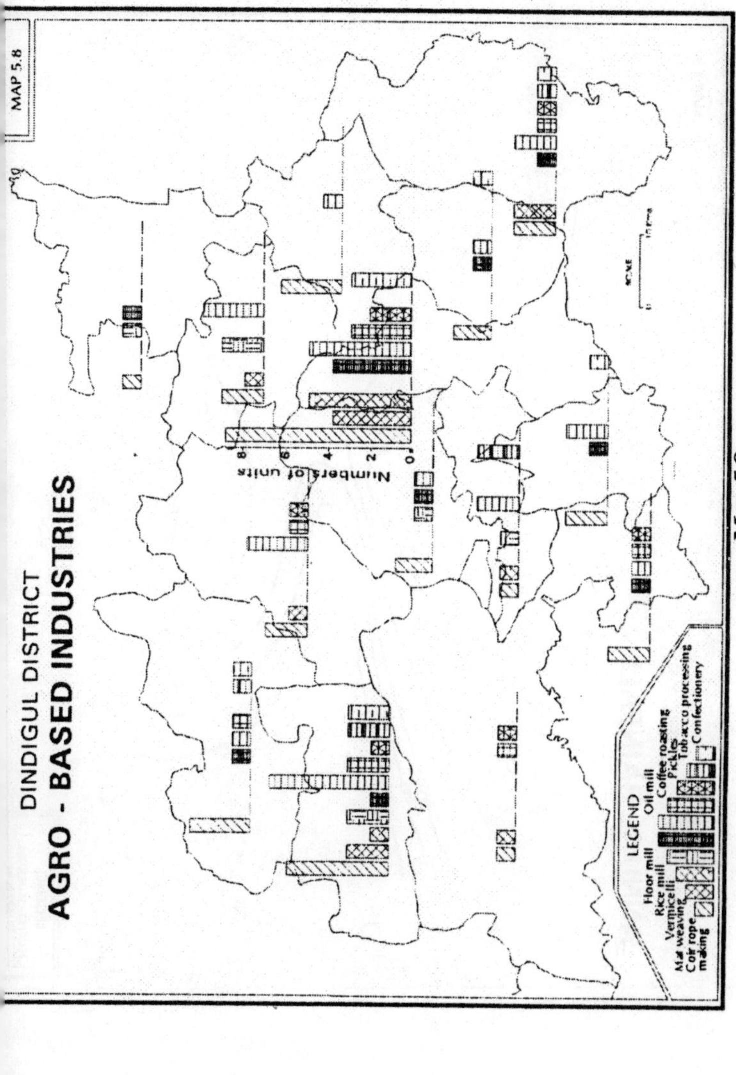
DINDIGUL DISTRICT
AGRO - BASED INDUSTRIES
MAP 5.8
Numbers of units
LEGEND
Oil mill
Rice mill
Vermicelli
Coffee roasting
Pickles
Tobacco processing
Confectionery

Map 5.8

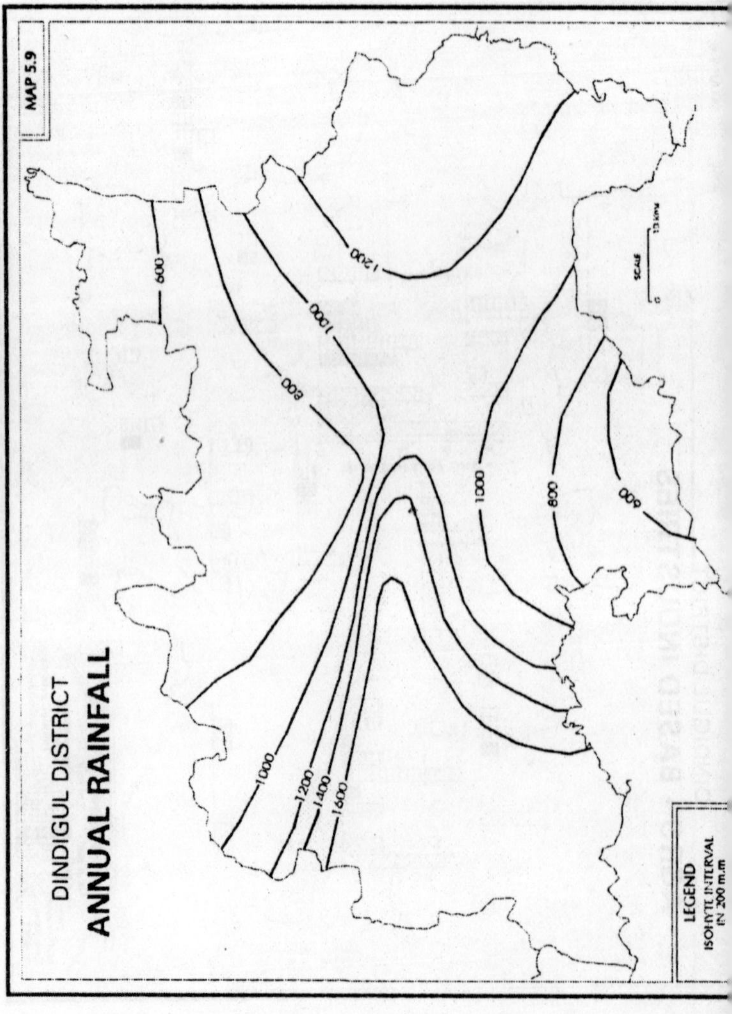
MAP 5.9
DINDIGUL DISTRICT
ANNUAL RAINFALL
LEGEND
ISOHYTE INTERVAL
IN 200 m.m
600
800
1000
1200
1400
1600
SCALE

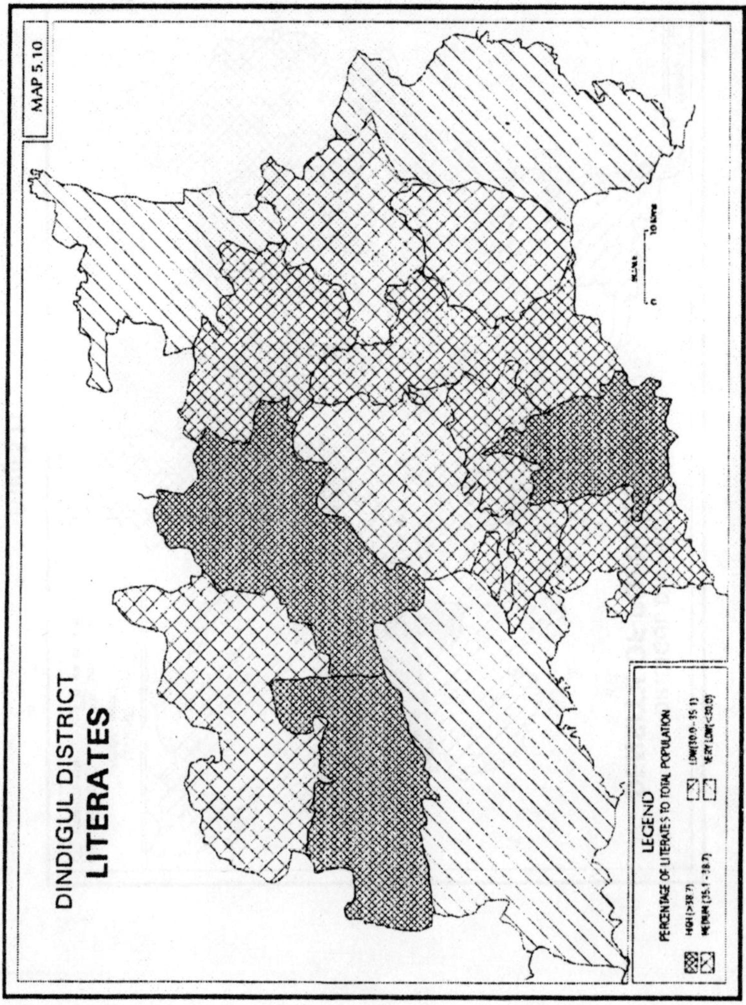
MAP 5.10
DINDIGUL DISTRICT
LITERATES
LEGEND
PERCENTAGE OF LITERATES TO TOTAL POPULATION
HIGH (>38.7)
MEDIUM (35.1 - 38.7)
LOW (30.0 - 35.1)
VERY LOW (<30.0)

Map 5.10

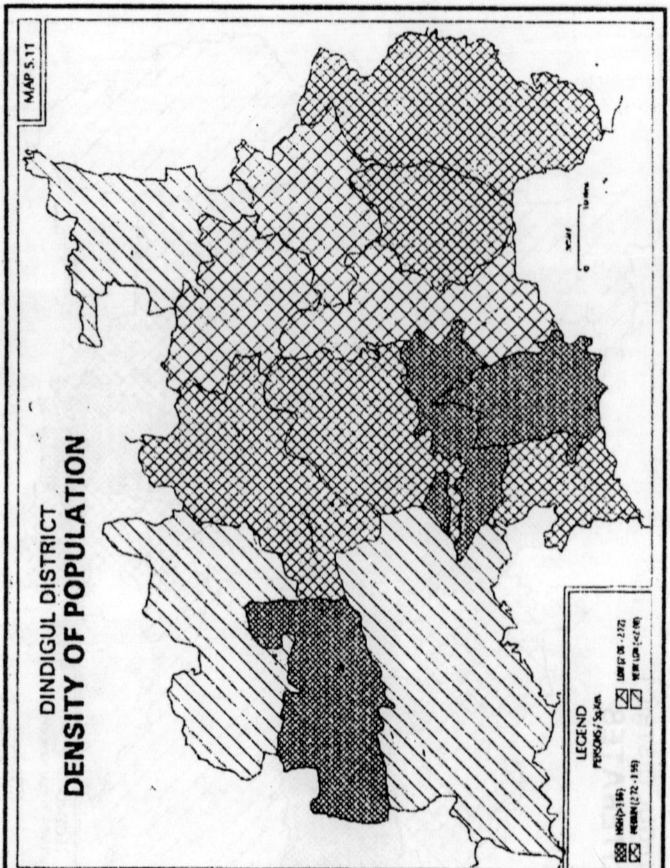
MAP 5.11
DINDIGUL DISTRICT
DENSITY OF POPULATION
LEGEND
PERSONS / Sq.KM.

Map 5.11

educational qualification is B.E., and at present works as an Executive Engneer in one of the Departments of Agriculture. He has put in 20 years of experience. As an Engineering student, he has learnt the preparation of maps for project formulation and use them till he evaluates the project. As part of his departmental operations, he has been preparing and using maps in his office. His views on the maps are presented here in his own words.

"I have been using maps in our departmental activities. If the maps are in colour it is easy to comprehend and appreciate them. As drawing of maps is a laborious work, I would like to use computers in preparing them. Therefore, I welcome the use of computers in map drawing. I realise the effectiveness of cartographic techniques in presenting large volume of data. I do see the value of maps as visual medium of communication from which I am able to get necessary information at a glance. There is a problem – how to select the cartographic technique. As the head of the office, I am not directly involved in the preparation of maps. My subordinates do that job. Therefore, it would be better, if our people have necessary skills in cartographic work. Alternatively, if there is an agency at the district to prepare and supply the maps, I would be very happy to use them in our work. As far as scale is concerned, we are interested in 1:50000 and field measurement maps. How can you help us in this area? I am also happy to see for the first time, region-wise distribution of the data like rainfall etc., instead of location-wise data. I hope this will be a pioneering effort towards supplying information to the officials in map form which will help them in quickening the process of decision-making. Will you give me copies of the maps on Drainage and Relief? asks the Official.

His views and observations show how relevant maps are in the daily routines of a development department.

(b) Respondent 'B' from potential map users : Respondent 'B' is an IAS Officer working for the development of rural areas. He is a Post Graduate with five years experience as an IAS Officer. He is in the present position for the past two years. He has got an exposure on maps and has used some maps too in his day to day activities. Though there is a map on his table, he uses it only for chalking out his tour programme.

He has the following views on the use of maps:

"I am using the map only for finalising my visits and not for any other departmental work, since maps required by me are not available even if they are available, they are not accessible. I have neither money for making maps nor I have a draughtsman. I personally feel, maps have a role to play in rural development work for instance, to increase the capacity of tanks and also to identify the ayakat area. For this purpose, I would like to have a map showing the location of tanks, their capacity and ayakat area. But such maps are not available. So far, I have seen but not used the maps published by the Department of Survey and Land Records. With regard to maps shown by you, I prefer colour maps. Today, I have seen a set of maps drawn by you using different techniques. Though I am able to understand the data presented in the maps, I like to know about the techniques of mapping and their characteristics. This would help me to instruct my subordinates to acquire or prepare maps as required by our work. Hence an exposure is a must for my subordinates. We need only block-wise maps showing activities of our department and also locally available resources in the paper size of your model map."

The above views and ideas of two district level officers have led the researcher to certain findings. Many of the officers who do not use maps, will use them if appropriate maps are made available to them. Even the few officials who have been using maps, have been unaware of the spatial dimension of the data. When informed of this unique aspect of maps, they tend to exhibit enthusiasm for the use of maps. Further, they would like to have some exposure to both map drawing and reading. They also prefer to give proper training and exposure to their subordinates in map drawing and reading. They expect maps to be handy and convenient in the size of model maps (*i.e.* 26 cms x 33 cms).

5.2.6 Perception Regarding Proposed Resource Atlas

The respondents were requested to give their perception of (a) level of data that could be used for the atlas, (b) scale of the atlas, (c) colour of maps, (d) the need for explanatory note for the map, and (e) the periodicity of updating the resource atlas.

A set of resource maps of Dindigul District comprising 49 selected maps (Appendix II) which had been prepared as part of this research work was presented to the respondents alongwith a list of 258 themes. The views of the respondents are presented below.

Table 5.5 : Perception of Respondents about various aspects of District Resource Atlas

Sl. No.	Aspects	Respondents in Percentage
(a)	Level of Data	
	Panchayat	15.0
	Revenue village	35.0
	Block	30.0
	Taluk	2.5
	Others (District)	17.5
(b)	Scale of the Map	
	1:250000	80.0
	1:50000	10.0
	1:16000	2.5
	Others	7.5
(c)	Colour of Maps	
	Colour	85.5
	Black and white	14.5
(d)	Explanatory Note	
	Required	52.5
	Not required	47.5

(a) Level of data: The data collected on various aspects shows that, data at the revenue village level is favoured by the largest percentage of respondents (35 per cent) followed by block (30 per cent) and *panchayat* (15 per cent).

(b) Scale of the maps: Of the total respondents, 80 per cent of them opted for the scale of 1:250000 while others asked for other scales (such as 1:7000, etc.).

(c) Colour of the maps: Most of them (85.5 per cent) have favoured colour maps, while 14.5 per cent are satisfied with black and white maps.

(d) Explanatory notes: With regard to explanatory notes, about 55 per cent of them want to have it for all maps in the resource atlas, while others are not very strict about it.

(e) Periodicity: Periodicity of updating the maps is crucial for the relevance and use of the maps. Undue delay in revision will be both expensive and unnecessary. Hence the need for determination of appropriate periodicity of the revision of maps. Views of the respondents on this question are furnished in Table 5.6.

Table 5.6 : Distribution of Map Users according to the Periodicity Suggested

Sl. No.	Periodicity (in years)	Respondents in Percentage
1.	1	57.5
2.	2	10.0
3.	3	2.5
4.	5	25.0
5.	10	5.0
	Total	100.0

It is clear from the above table that, 25 per cent of the respondents required an updated atlas once in 5 years whereas 57.5 per cent of them need it once a year.

It is made known by the analysis that, more than half of them required updated version once a year, while one-fourth required it in 5 years.

5.2.7 Themes Selected by the Respondents

After going through the list of 258 themes presented by the researcher (including 49 maps) users selected 100 themes. The list of themes found useful by them is furnished in Appendix IX.

5.2.8 New Themes Required by the Respondents

It must however be pointed out that 65 per cent of the respondents expressed their need for certain maps which were not included in the list of themes presented by the researcher to them. Particulars of the theme of maps as sought by these departments are given in Table 5.7.

Table 5.7 : New Themes Required by the Respondents

Sl. No.	Theme of the Maps	Department Seeking the Map
1.	SC/ST Land holding (size and number)	Agri. Engineering (Soil Conservation) DRDA
2.	Location of check dams and percolation ponds	Public Works Department, DRDA

(Contd.)

Table 5.7 *(Contd.)*

Sl. No.	Theme of the Maps	Department Seeking the Map
3.	Location of public buildings	Public Works Department
4.	Rural works of Panchayat Union	Highways and Rural Works
5.	Traffic census	Highways and Rural Works
6.	Porosity of soil	TWAD (RWS)
7.	Texture of soil	TWAD (RWS)
8.	Flurosis prone area and intensity	TWAD (RWS)
9.	Accessibility of Primary Health Centre and Sub Centres	Public Health
10.	Prevailing health conditions*	Public Health
11.	Population served by PHCs	Public Health
12.	Disease pattern — seasonality and occurrence	Public Health
13.	Incidence of diseases	Public Health
14.	Pattern of houses (Block-wise maps)*	Public Health
15.	Disease-affected houses (Block-wise maps)*	Public Health
16.	Source of infection (Block-wise maps)*	Public Health
17.	Location of harijan colony (Block-wise maps)*	Public Health
18.	Age-structure of children population	Public Health
19.	Polio affected children	Public Health
20.	Chief Minister's Noon Meal Centres and ownership of buildings	Public Health
21.	Community toilets*	Public Health
22.	Accident prone areas	Public Health
23.	Bus routes	RTO
24.	Forest division map (Showing range, section and beat with Head Quarters)	Forest
25.	Ownership of forest with forest roads	Forest

(Contd.)

Table 5.7 *(Contd.)*

Sl. No.	Theme of the Maps	Department Seeking the Map
26.	Disease map	Agriculture (Training and Visit)
27.	Distribution of seed under special food grain production programme	Agriculture
28.	AAO's head quarters with days of visit to villages	Agriculture (Training and Visit)
29.	Production of fodder	Animal Husbandry
30.	Incidence of diseases	Animal Husbandry
31.	Prevalence of diseases	Animal Husbandry
32.	Water spread area of Tank	Fisheries, *Panchayat* Development and DRDA
33.	Ownership of tank	Fisheries, *Panchayat* Development and DRDA
34.	Fishing rights	Fisheries, *Panchayat* Development and DRDA
35.	Status of tank	Fisheries, *Panchayat* Development and DRDA
36.	Progressive coverage under IRDP	Lead Bank
37.	Progressive coverage under biogas	Lead Bank
38.	Progressive coverage under self-employed youth	Lead Bank
39.	Accessibility of schools*	Education
40.	Availability of Infrastructure facilities in schools (lab etc.)	Education
41.	Adult illiterates as percent of total population*	Non-formal and Adult Education
42.	Availability of jobs for illiterates*	Non-formal and Adult Education
43.	Utilization pattern of resources: Water, Fuel (firewood and power)*	Non-formal and Adult Education
44.	Location of mahalir mandram*	Social Welfare

(Contd.)

Table 5.7 *(Contd.)*

Sl. No.	Theme of the Maps	Department Seeking the Map
45.	Location of tailoring society*	Social Welfare
46.	Concentration of freed bonded labourers*	Adi Dravida Welfare DRDA
47.	SC & ST landless agricultural labourers	Adi Dravida Welfare
48.	Industries run by SC/ST*	Adi Dravida Welfare
49.	Percentage of couples protected to total eligible couples	Family Welfare
50.	Couple protection rate among literates	Family Welfare
51.	Female population (15-25 Yrs) percentage of total population	ICDS
52.	Female infanticide*	ICDS
53.	Female diseases*	ICDS
54.	Nutritional status of women*	ICDS
55.	Female — health care facilities	ICDS
56.	Female — mortality	ICDS
57.	Female - morbidity	ICDS
58.	Problem children*	ICDS
59.	Post natal care*	ICDS
60.	Prospective locations for percolation ponds*	*Panchayat* Development
61.	Location of town *panchayat*	Town *Panchayat*
62.	Location of hand pumps, power pumps, ground level reservoir and over head tanks in town *panchayats**	Town *Panchayat*
63	Group houses of SC/ST*	Town *Panchayat*
64.	Sewage system*	Town *Panchayat*
65.	Accessibility of public distribution system	District Supply Officer
66.	Location and number of children fed in balwadies	ICDS
67.	Union school buildings (Elementary to Higher Secondary)	DRDA
68.	*Panchayat* office buildings	DRDA

(Contd.)

Table 5.7 *(Contd.)*

Sl. No.	Theme of the Maps	Department Seeking the Map
69.	Area irrigated by *panchayat* tanks to the registered area	DRDA
70.	Land holding of freed bonded labourers*	DRDA
71.	Location of hostel	Backward class and Minorities Welfare
72.	Location of dhobi khanas	Backward class and Minorities Welfare
73.	Location of boarding grant getting hostel	Backward class and Minorities Welfare
74.	Government owned and rented hostels	Backward class and Minorities Welfare
75.	Availability of biomass	Tamil Nadu State Electricity Board

* Themes discorded by map designers.

5.3 PERCEPTION OF MAP DESIGNERS

As discussed earlier a list of maps and map themes were given to the map user to study their perception of different aspects of the maps. Similarly, the same list containing 258 map themes and also another list of 75 themes required by the map users were sent to 63 map designers for the purpose of collecting their views on themes of map scale, periodicity and level of data to be employed in the preparation of resource atlas for district level planning. Of the total respondents (63), only 19 of them completed the schedule with all entries and mailed the same to the researcher. Six respondents informed the researcher that they had no comments on the list sent to them. This includes the National Institute of Rural Development (NIRD), who has sent to the researcher, three sets of maps which were used by them in their district planning exercise for Aurangabad and Nasik districts of Maharashtra State and Raichur district of Karnataka State.

(a) Level of data: The particulars regarding the most appropriate level of data suggested by the map designers are furnished in the Table 5.8.

Table 5.8 : Level of Data as Suggested by Map Designers

Sl. No.	Level of Data	Respondents in Percentage
1.	Revenue Village	53
2.	Block	16
3.	Taluk	5.3
4.	Not specified	26.3
	Total	100.0

From the table it is clear that, 53 per cent and 16 per cent suggested the revenue village level and block level data respectively as the level of data for drawing maps for district level planning. It is evident from the analysis, that majority of the respondents suggested revenue village and block level data suitable for resource atlas.

(b) Scale of the Atlas: The data on the scale of the atlas suggested by the map designers are presented in Table 5.9.

Table 5.9 : Scale of the Atlas as suggested by the Map Designers

Sl. No.	Scale of Atlas	Respondents in Percentage
1.	1:500000	5.3
2.	1:250000	37
3.	1:50000	11
4.	1:10000	5
5.	Not specified	42.1
	Total	100.0

With regard to the scale of the maps, 37 per cent of them have suggested 1:250000 scale as the suitable scale for the district level resource atlas, while 11 per cent required only 1:50000 scale maps. 5 per cent each has asked for 1:500000 and 1:10000 respectively. The remaining 42 per cent have not mentioned the scale. Thus it may be inferred from the analysis that, the most appropriate scale as suggested by the map designers is 1:250000.

(c) Periodicity: The importance of the periodicity of the updating of maps has already been highlighted in the treatment of the perceptions of the map users. Views of the map designers in respect of optimum time-lag of revision are furnished in Table 5.10.

Table 5.10 Distribution of Respondents According to the Periodicity of Revision

Sl. No.	Periodicity (in years)	Respondents in Percentage
1.	Less than 5	16
2.	5	42
3.	10	11
4.	Not specified	32
	Total	100.0

As seen from the above table 42 per cent of them wanted to update the proposed atlas once in 5 years, while 11 per cent once in 10 years. Another 16 per cent required the updated version in less than 5 years while the remaining 32 per cent have not specified the period of updating.

It may be inferred from the above analysis that the majority of the map designers preferred to update the resource atlas for district planning once in 5 years.

5.3.1 Maps Selected by the Map Designers

All the 258 themes prepared by the researcher were accepted by the map designers. Of the 75 themes required by the map users, 50 themes selected by the map designers are presented in Table 5.7.

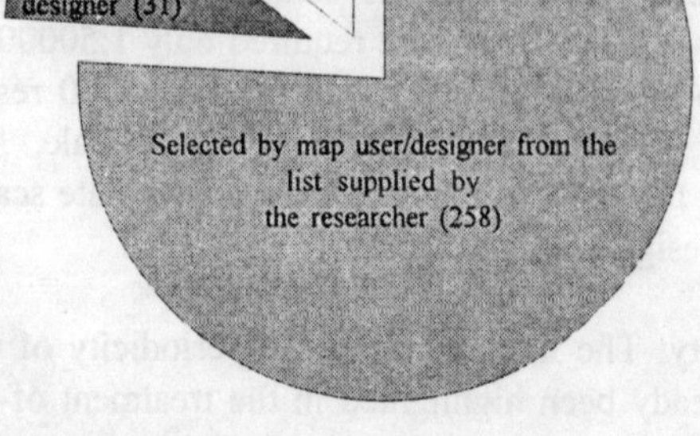

Fig 5.1 : Sources of Themes for District Resource Atlas

It is, therefore, proposed to include all the 258 themes in the proposed district resource atlas, even though 158 themes were not selected by the district level users. As already pointed out, the practice of using maps is yet to find its due place at the district level. It is hoped that with the passage of time, those of the 158 themes which are not now required by them will become useful as district level authorities become familiar with the use of maps. Similarly among 75 suggested by the users 50 themes which were accepted by designers will be added to the above list.

5.3.2 New Themes Suggested by the Map Designer

Most of the designers have suggested new themes for district resource atlas (Table 5.11).

Thus, the proposed district resource atlas will contain 339 themes consisting of 258 themes compiled by the researcher, 50 themes required by the users and endorsed by the map designers and 31 themes suggested by the map designers (Fig. 5.1).

5.4 DISCUSSION AND OUTCOME

5.4.1 Level of Data

It has been noticed earlier that 35 per cent of the map users would need data at revenue village level, 15 per cent at *panchayat* level and 30 per cent at block level. Among the designers, 52 per cent of them suggest data at the revenue village level and 16 at the block level with none opting for *panchayat* level data. Decision on the level of data to be presented in the map involves a conflict which is difficult to resolve. Logics of spatial planning warrants as lower level of data as possible-say individual village or hamlet whereas the feasibility of accommodating data in a compact sheet of map to be visually appreciated by the map user calls for a level of data at a far higher plane. It is true that most of the maps in Medak District Atlas, present village level data (Table 4.3) and a perusal of these maps however has convinced the researcher that this level of data is likely to render maps incomprehensible and conglomerate with little appeal to the visual understanding of the users.

Again there is a conflict between revenue village and *panchayat* with different boundaries. It is difficult to choose one over the other because revenue village constitutes the basic unit of revenue administration, while *panchayat* is the basic unit for rural development programme. However, the boundaries of revenue villages and *panchayats* are coterminous with block boundaries. It is therefore, proposed that block may be the level of data for district level planning. It may be recalled that district level planning will be decentralised at block level which becomes the lowest planning unit. Hence, the choice of block as the level of data for maps to be used for district level planning. In this context reference must be made to 73rd constitutional amendment

Table 5.11 : List of Maps Suggested by the Map Designers

Sl. No.	Themes of Map
1.	General disease area
2.	Consumption pattern of petrol
3.	Consumption pattern of diesel
4.	Consumption pattern of kerosene
5.	Intensity of irrigation
6.	Distribution of unemployed people
7.	Distribution of skilled unemployed people
8.	Distribution of unskilled unemployed people
9.	Distribution of village industries
10.	Minor irrigation scheme
11.	Depth of water table (pre-monsoon)
12.	Depth of water table (post-monsoon)
13.	Normal temperature
14.	Use of major fertilizers (spatial distribution)
15.	Ground water pollution
16.	Changes in the density of population 1961-71
17.	People's choice of centres at first level of functional hierarchy
18.	People's choice of centres at second level of hierarchy
19.	People's choice of centres at third level of hierarchy
20.	Growth centres, Service centres and Central villages
21.	Centres and their hinterlands — First level
22.	Centres and their hinterlands — Second level
23.	Centres and their hinterlands — Third level
24.	Central villages and their hinterlands
25.	Service villages and their hinterlands
26.	Growth centres and their hinterlands
27.	Single teacher school
28.	Households below poverty line
29.	Capacity and utilization of roads
30.	Commodity flow by roads
31.	Over utilized roads

giving new strength to *Panchayati Raj*, especially through mandatory election within six months after the expiry or supersession of elected body and quinquennial finance commission for devolution of resources.

Therefore, in future village *panchayat* will have to play a vital role as the basic unit of decentralised planning. In the emerging situation, block-wise maps with *panchayat* level data may be needed for planning at the level of block.

5.4.2 Periodicity of Updating

The maps will have to be periodically updated in order to serve the purpose of aiding the planning process. Among the users, 58 per cent favoured annual revision of maps because they want to present in maps annual progress made in the implementation of plans. About 25 per cent would like to have revision of maps once in five years. Among designers, 42 per cent suggest five years as the period for revision of maps, while about 16 per cent want shorter period for revision. The periodicity of revision of maps will have to be determined with reference to the requirement of purpose, availability of facilities and cost. Plans are drawn for a period of five years. Therefore, revision of maps once in five years will serve the purpose of presenting the achievements of the preceding five year plan and preparing the ground for the current plan.

Survey of the perception of map users and map designers, incontinuation of the exercise initated in chapter IV, has facilitated finalisation of the list of maps, as required for district level planning. As part of this endeavour, it has been possible to identify the map themes, technique of mapping, scale, level of data and periodicity of updating as appropriate to district level planning. On the basis of the above findings, the following chapter will be devoted to the formulation of model district resource atlas.

6

A MODEL RESOURCE ATLAS OF DINDIGUL DISTRICT, TAMIL NADU

6.1 INTRODUCTION

The present chapter is addressed to preparation of a model District Resource Atlas and it will be based on the findings arrived in Chapter IV and V.

6.2 A MODEL DISTRICT RESOURCE ATLAS – GENERALISED VIEW

6.2.1 Themes of Map

In the present investigation, map themes required for district level planning were identified in two stages. In Chapter 4, evaluation of four atlases and perusal of guidelines of the Planning Commission for district level planning yielded 258 themes. Study of the views of the map users and designers in Chapter 5 contributed 81 themes. Thus the researcher has identified 339 themes. (Table 6.1 and Fig. 6.1).

6.2.2 Techniques of Mapping

Method of presentation of data is very important, if the maps are to serve the purpose of visual appreciation by the decision makers. The mapping technique has already been decided for 258 maps in Chapter IV. The researcher has identified appropriate methods for the remaining 81 themes in column No. 3 of Table 6.1.

6.2.3 Scale of Maps

Evaluation of Medak district atlas has shown that district level maps have been drawn in the scale of 1:500000. However, it was noticed in

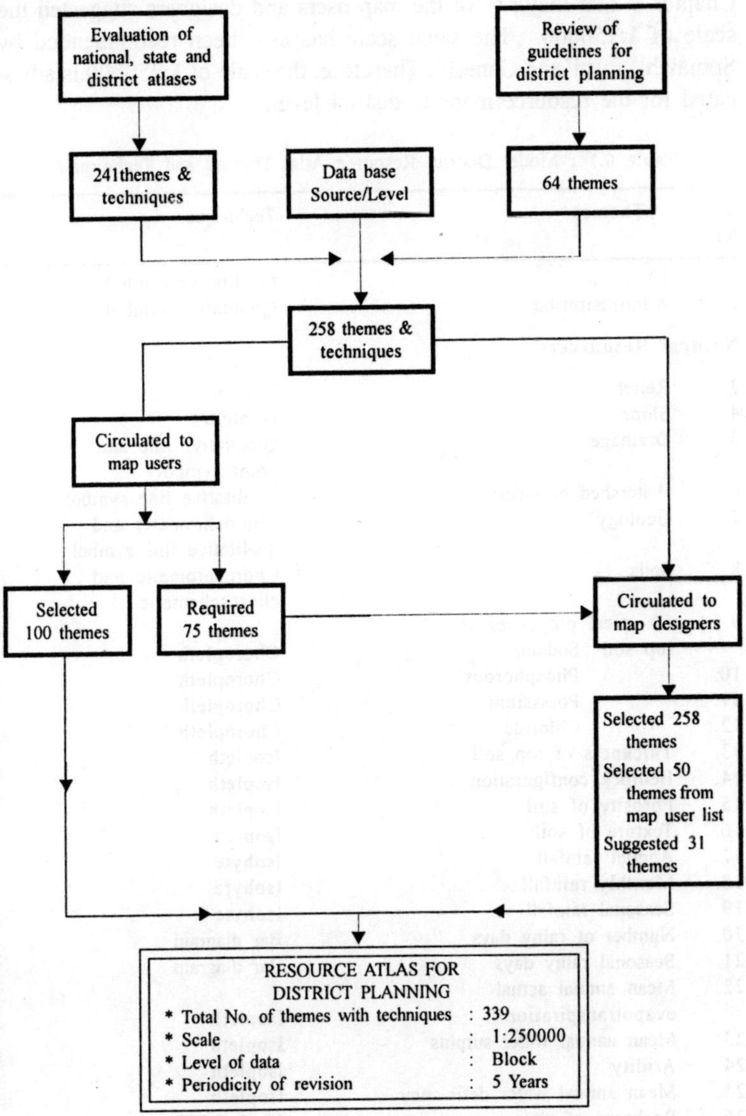

Fig. 6.1 : Methodology followed in Designing a Resource Atlas for District Level Planning

Chapter V that majority of the map users and designers suggested the scale of 1:250000. The same scale has also been recommended by Somaiah[1], Joshi[2] and Ganesh[3]. Therefore, the scale of 1:250000 is advocated for the resource maps at district level.

Table 6.1 : Model District Resource Atlas-Themes and Techniques

Sl. No.	Themes	Techniques
1.	Location	Qualitative symbol
2.	Administration	Qualitative symbol
Natural Resources		
3.	Relief	Contour
4.	Slope	Isopleth
5.	Drainage	Qualitative line and point symbol
6.	Watershed boundary	Qualitative line symbol
7.	Geology	Chorochromatic and qualitative line symbol
8.	Soils	Chorochromatic and choroschematic
9.	Chemical properties of top soil : Sodium	Choropleth
10.	Phosphorous	Choropleth
11.	Potassium	Choropleth
12.	Chloride	Choropleth
13.	Thickness of top soil	Isopleth
14.	Bedrock configuration	Isopleth
15.	Porosity of soil	Isopleth
16.	Texture of soil	Isopleth
17.	Annual rainfall	Isohyte
18.	Monthly rainfall	Isohyte
19.	Seasonal rainfall	Isohyte
20.	Number of rainy days	Bar diagram
21.	Seasonal rainy days	Bar diagram
22.	Mean annual actual evapotranspiration	Isopleth
23.	Mean annual water surplus	Isopleth
24.	Aridity	Isopleth
25.	Mean annual water deficiency	Isopleth
26.	Recharge of water	Isopleth
27.	Water run-off	Isopleth
28.	Variability of rainfall	Isopleth
29.	Annual temperature	Isotherm
30.	Monthly temperature	Isotherm

(Contd.)

Table 6.1 *(Contd.)*

Sl. No.	Themes	Techniques
31.	Normal temperature	Isotherm
32.	Wind speed and direction	Wind rose
33.	Natural vegetation	Chorochromatic
34.	Forest	Chorochromatic
35.	Ownership of forest with forest roads	Chorochromatic and qualitative line symbol
36.	Forest divisions with range, section and beat including headquarters	Qualitative line and point symbol
37.	Forest revenue	Proportionate squares
38.	Forest utilization: Fuelwood	Bar diagram
39.	Pulp and other Indian Wood	Bar diagram
40.	Plywood	Bar diagram
41.	Minor forest produce	Bar diagram
42.	Natural regions	Chorochromatic
43.	Bio-climatic Regions	Chorochromatic
Ground Water/Irrigation		
44.	Hydro-geology	Chorochromatic
45.	Depth of water table	Isopleth
46.	Depth of water table (pre-monsoon)	Isopleth
47.	Depth of water table (post-monsoon)	Isopleth
48.	Ground water quality	Isopleth
49.	Ground water potential	Chorochromatic
50.	Sources of irrigation including location of dams with command area	Chorochromatic, Qualitative line and point symbol
51.	Location of check dams and percolation ponds	Qualitative point symbol
52.	Minor irrigation schemes	Qualitative point symbol
53.	Extent of irrigation	Pie-diagram
54.	Irrigated area : Tank	Choropleth
55.	Well	Choropleth
56.	Canal	Choropleth
57.	Other sources	Choropleth
58.	Net irrigated area	Choropleth
59.	Total irrigated area	Choropleth
60.	Changes in the extent of irrigated area	Choropleth
61.	Intensity of irrigation	Choropleth
62.	Area irrigated by *panchayat* tanks to registered area	Choropleth
Land Use/Agriculture		
63.	Land use	Chorochromatic

(Contd.)

Table 6.1 *(Contd.)*

Sl. No.	Themes	Techniques
64.	Land capability	Chorochromatic
65.	Percentage of barren and uncultivable land	Choropleth
66.	Percentage of cultivable waste land	Choropleth
67.	Percentage of fallows other than current fallows	Choropleth
68.	Percentage of current fallows	Choropleth
69.	Percentage of permanent pastures and grazing lands	Choropleth
70.	Percentage of area under non-agricultural uses	Choropleth
71.	Percentage of area under miscellaneous tree crops	Choropleth
72.	Percentage of net sown area	Choropleth
73.	Percentage of area under forest	Choropleth
74.	Distribution of wastelands	Chorochromatic
75.	Cropping pattern	Pie-diagram
76.	Area under Kharif Dry crops	Pie-diagram
77.	Area under Rabi crops	Pie-diagram
78.	Area and yield of principal food crops	Choropleth
79.	Area and yield of principal commercial crops	Choropleth
80.	Area under Rice	Choropleth
81.	Area under Bajra	Choropleth
82.	Area under Jowar	Choropleth
83.	Area under Maize	Choropleth
84.	Area under Sorghum	Choropleth
85.	Area under Pulses	Choropleth
86.	Area under Banana	Choropleth
87.	Area under Other plantain	Choropleth
88.	Area under Tapioca	Choropleth
89.	Area under Groundnut	Choropleth
90.	Area under Sesame	Choropleth
91.	Area under Castor	Choropleth
92.	Area under Coconut	Choropleth
93.	Area under Other oilseeds	Choropleth
94.	Area under Sugarcane	Choropleth
95.	Area under Cotton	Choropleth
96.	Area under Tea	Choropleth
97.	Area under Coffee	Choropleth
98.	Area under Pepper	Choropleth
99.	Area under Betelnut	Choropleth
100.	Area under Rubber	Choropleth

(Contd.)

Table 6.1 *(Contd.)*

Sl. No.	Themes	Techniques
101.	Yield of Rice	Choropleth
102.	Yield of Bajra	Choropleth
103.	Yield of Jowar	Choropleth
104.	Yield of Maize	Choropleth
105.	Yield of Sorghum	Choropleth
106.	Yield of Pulses	Choropleth
107.	Yield of Banana	Choropleth
108.	Yield of Other plantain	Choropleth
109.	Yield of Tapioca	Choropleth
110.	Yield of Groundnut	Choropleth
111.	Yield of Sesame	Choropleth
112.	Yield of Castor	Choropleth
113.	Yield of Coconut	Choropleth
114.	Yield of Other oilseeds	Choropleth
115.	Yield of Sugarcane	Choropleth
116.	Yield of Cotton	Choropleth
117.	Yield of Tea	Choropleth
118.	Yield of Coffee	Choropleth
119.	Yield of Pepper	Choropleth
120.	Yield of Betelnut	Choropleth
121.	Yield of Rubber	Choropleth
122.	Changes in the area under : Rice	Choropleth
123.	Changes in the area under : Sorghum	Choropleth
124.	Changes in the area under : Bajra	Choropleth
125.	Changes in the area under : Pulses	Choropleth
126.	Changes in the area under : Sugarcane	Choropleth
127.	Changes in the area under : Cotton	Choropleth
128.	Changes in the area under : Groundnut	Choropleth
129.	Changes in the area under : Other oilseeds	Choropleth
130.	Area sown more than once	Choropleth
131.	Intensity of cropping	Choropleth
132.	Man-land ratio	Choropleth
133.	Disease map (Pest Calendar)	Diagram
134.	State seed farms	Qualitative point symbol
135.	Storage godowns	Point symbol
136.	Distribution of seed under special food grain production programme	Bar diagram

(Contd.)

Table 6.1 *(Contd.)*

Sl. No.	Themes	Techniques
137.	Seed and fertilizer shops	Qualitative point symbol
138.	Use of important fertilizer (spatial distribution)	Proportionate point symbol
139.	Markets	Qualitative point symbol
140.	Agricultural credit societies	Qualitative point symbol
141.	Asst. Agricultural officer/serving area headquarters with days of visit	Qualitative line, Point and area symbol
142.	Arrival of grains to markets (quantity & value)	Circle
143.	Grains: Estimated arrival and value	Circle
144.	Crop region	Chorochromatic and choroschematic
145.	Landholdings of small and marginal farmers	Pie-diagram
146.	Landholdings: Area	Pie-diagram
147.	Number	Pie-diagram
148.	Landholdings of tenant cultivators : Area	Pie-diagram
149.	Number	Pie-diagram
150.	SC & ST landholding : Area	Pie-diagram
151.	Number	Pie-diagram
152.	Tractors and tractor operated implements	Choropleth
153.	Bullock and manually operated implements	Choropleth
154.	Oil engines, electrical pumpset and sprayers	Choropleth
155.	Sugarcane crushers, oil ghannies and thrushers	Choropleth
Animal Husbandry		
156.	Percentage distribution of livestock	Pie-diagram
157.	Milch animal	Choropleth
158.	Dairy projects	Circles
159.	Milk routes	Qualitative line symbol
160.	Poultry farms	Quantitative point symbol
161.	Distribution of poultry population	Quantitative point symbol
162.	Density of poultry population	Choropleth
163.	Availability of poultry/ 1000 persons	Choropleth
164.	Marine fisheries	Pie-diagram
165.	Inland fisheries	Pie-diagram
166.	Area of water spread in tanks	Choropleth

(Contd.)

Table 6.1 *(Contd.)*

Sl. No.	Themes	Techniques
167.	Ownership of tanks	Qualitative point and choroschematic
168.	Fishing rights of tanks	Choroschematic
169.	Status of tanks	Qualitative area symbol
170.	Fishing equipments	Bar diagram
171.	Fisheries infrastructure	Qualitative point symbol
172.	Market for fish	Qualitative point symbol
173.	Market for livestock	Qualitative point symbol
174.	Production of fodder	Qualitative point symbol
175.	Arrival of livestock to market (quantity and value)	Circles
176.	Incidence of diseases	Bar diagram
177.	Prevalence of diseases	Bar diagram
	Population	
178.	Growth of population 1981-91	Choropleth
179.	Growth of population 1971-81	Choropleth
180.	Growth of population 1951-91	Choropleth
181.	Density of population 1981	Choropleth
182.	Density of population 1991	Choropleth
183.	Changes in density of population 1981-91	Choropleth
184.	Distribution of population	Quantitative point and circle
185.	Migration	Pie-diagram
186.	Total Scheduled Castes population 1981	Choropleth
187.	Total Scheduled Castes population 1991	Choropleth
188.	Total Scheduled Tribes population 1991	Choropleth
189.	Sex-ratio	Choropleth
190.	Decadal variation of sex-ratio	Choropleth
191.	Literates	Choropleth
192.	Male literates	Choropleth
193.	Decadal variation in literacy	Choropleth
194.	Birth, death and infant mortality rates	Located graph
195.	Female mortality rates	Located graph
196.	Female morbidity rates	Located graph
197.	Rural urban composition	Pie-diagram
198.	Age-structure of population	Pyramidal diagram
199.	Occupational structure	Pie-diagram
200.	Percentage of workers	Choropleth
201.	Percentage of non-workers	Choropleth

(Contd.)

Table 6.1 *(Contd.)*

Sl. No.	Themes	Techniques
202.	Distribution of educated, unemployed persons	Choropleth
203.	Distribution of skilled, unemployed people	Choropleth
204.	Distribution of unskilled, unemployed people	Choropleth
205.	Percentage of cultivators	Choropleth
206.	Percentage of male cultivators	Choropleth
207.	Percentage of agricultural labourers	Choropleth
208.	Percentage of SC/ST agricultural labourers	Choropleth
209.	Percentage of male agricultural labourers	Choropleth
210.	Percentage of workers in household industries	Choropleth
211.	Percentage of male workers in household industries	Choropleth
212.	Percentage of marginal workers	Choropleth
213.	Percentage of male marginal workers	Choropleth
214.	Percentage of main workers	Choropleth
215.	Percentage of male main workers	Choropleth
216.	Percentage of other workers	Choropleth
217.	Disease pattern — seasonality, occurrence	Choropleth and graph
218.	Incidence of diseases	Circle
219.	General disease area	Chorochromatic
220.	Age-structure of children population	Pyramidal bar
221.	Polio affected children	Circle
222.	Percentage of couples protected to total eligible couples	Choropleth
223.	Couple protection rate among literates	Choropleth
224.	Percentage of female population (15-25 years)	Choropleth
225.	Distribution of urban and rural settlements	Circle
226.	Functional hierarchy of urban centres	Circle
227.	Peoples' choice of centres at first level of functional hierarchy	Ray diagram
228.	Peoples'. choice of centres at second level of functional hierarchy	Ray diagram

(Contd.)

Table 6.1 *(Contd.)*

Sl. No.	Themes	Techniques
229.	Peoples' choice of centres at third level of functional hierarchy	Ray diagram
230.	Growth centres, service centres and central villages	Qualitative point symbol
231.	Centres and their hinterlands — first level	Qualitative point and area symbol
232.	Centres and their hinterlands — second level	Qualitative point and area symbol
233.	Centres and their hinterlands — third level	Qualitative point and area symbol
234.	Central villages and their hinterlands	Qualitative point and area symbol
235.	Service villages and their hinterlands	Qualitative point and area symbol
236.	Growth centres and their hinterlands	Qualitative point and area symbol
Rural Development		
237.	Construction of buildings under National Rural Employment Programme	Qualitative point symbol
238.	Supply of drinking water: Source and quality of water	Qualitative point symbol and Pie-diagram
239.	Drinking water problem villages	Qualitative point symbol
240.	Rural works of *panchayat* Union	Qualitative point symbol
241.	Location of town *panchayat*	Qualitative point symbol
242.	*Panchayat* office buildings	Qualitative point symbol
243.	Location of public buildings	Qualitative point symbol
244.	Rural housing programme	Circle
245.	Housing for Scheduled Castes	Circle
246.	Housing for Scheduled Tribes	Circle
247.	Housing for Backward classes	Circle
248.	Housing for economically backward classes	Circle
249.	IRDP loans advanced to beneficiaries under various schemes	Bar diagram
250.	Households below poverty line	Bar diagram
251.	IRDP : Progressive coverage	Bar diagram
252.	Progressive coverage : Bio-gas	Bar diagram
253.	Self employed youth	Bar diagram
254.	Location of dhobi khannas	Qualitative point symbol

(Contd.)

Table 6.1 *(Contd.)*

Sl. No.	Themes	Techniques
Industries		
255.	Power grid with transformers	Qualitative line and point symbol
256.	Villages to be electrified	Qualitative point symbol
257.	Power generation	Circle
258.	Power consumption	Circle
259.	Location of mines	Qualitative point symbol
260.	Production of minerals	Circle
261.	Large and medium scale industries: Workers employed and capital invested	Pie-diagram
262.	Distribution of small scale industries	Qualitative point symbol
263.	Distribution of village industries	Qualitative point symbol
264.	Agro-based industries: No. of units and workers	Circle
265.	Capital invested and value of output	Circle
266.	Agri. implements manufacturing industries: No. of units and workers	Circle
267.	Capital invested and value of output	Circle
268.	Wood based industries : No. of units and workers	Circle
269.	Capital invested and value of output	Circle
270.	Rubber & Leather industries : No. of units and workers	Circle
271.	Capital invested and value of output	Circle
272.	Chemical Industries : No. of units and workers	Circle
273.	Capital invested and value of output	Circle
274.	Ceramic & Building materials industries: No. of units and workers	Circle
275.	Capital invested and value of output	Circle
276.	Engineering industries : No. of units and workers	Circle
277.	Capital invested and value of output	Circle

(Contd.)

Table 6.1 *(Contd.)*

Sl. No.	Themes	Techniques
278.	Electrical and allied industries : No. of units and workers	Circle
279.	Capital invested and value of output	Circle
280.	Printing, binding and allied industries : No. of units and workers	Circle
281.	Capital invested and value of output	Circle
282.	Miscellaneous industries:No.of units and workers	Circle
283.	Capital invested and value of output	Circle
284.	Location of industrial estates	Qualitative point symbol
285.	Proposed large and medium scale industries	Qualitative point symbol
286.	Tourism	Qualitative point and line symbol
287.	Wild life and sanctuaries	Qualitative point and line symbol
288.	Places of religious and historical importance	Qualitative point and line symbol
Infrastructure Facilities		
289.	Roads and railways	Qualitative line symbol
290.	Inland waterways	Qualitative line symbol
291.	Airways and sea routes	Qualitative line symbol
292.	Bus routes	Qualitative line symbol
293.	Bus traffic flow	Quantitative line symbol
294.	Traffic census	Quantitative line symbol
295.	Capacity and utilization of roads	Quantitative line symbol
296.	Over utilized roads	Quantitative line symbol
297.	Commodity flow of roads	Quantitative line symbol
298.	Accident prone areas	Qualitative line symbol
299.	Centres of education	Qualitative point symbol
300.	School education districts	Qualitative point symbol
301.	Single teacher school	Qualitative point symbol
302.	Students strength of centres of education	Circle
303.	Location and number of children fed in balwades	Qualitative point symbol and circle
304.	Union school buildings	Qualitative point symbol
305.	Chief Minister's noon meal centres and ownership of buildings	Qualitative point symbol and Choroschematic

(Contd.)

Table 6.1 *(Contd.)*

Sl. No.	Themes	Techniques
306.	Location of backward class hostels	Qualitative point symbol
307.	Boarding grant getting hostels	Qualitative point symbol
308.	Government owned/rented hostels	Qualitative point symbol
309.	Banks	Qualitative point symbol
310.	Service area of the banks	Qualitative area symbol
311.	Population served per bank	Choropleth
312.	Telegraph grid	Qualitative point and line symbol
313.	Road accessibility	Isopleth
314.	Road availability	Choropleth
315.	Road density	Choropleth
316.	Railway accessibility	Isopleth
317.	Post and telegraph offices	Qualitative point symbol
318.	Population served by post office	Choropleth
319.	Hospital	Qualitative point symbol
320.	Female health care facilities	Qualitative point symbol
321.	Primary health centres	Qualitative point symbol
322.	Accessibility of primary health centres and sub-centres	Isopleth and Qualitative point symbol
323.	Population served by primary health centres	Choropleth
324.	Veterinary facilities	Qualitative point symbol
325.	Accessibility of public distribution shops	Isopleth and qualitative point symbol
326.	Consumption pattern of kerosene	Circle
327.	Consumption pattern of diesel	Circle
328.	Consumption pattern of petrol	Circle
329.	Problem areas — Areas liable to Flooding	Chorochromatic
330.	Water-logging	Chorochromatic
331.	Soil erosion	Chorochromatic
332.	Alkalinity	Chorochromatic
333.	Recurrent drought	Chorochromatic
334.	Ground water pollution	Chorochromatic
335.	Flourosis prone area and intensity	Chorochromatic
	Regions and Development	
336.	Levels of agricultural development	Choropleth
337.	Levels of industrial development	Choropleth

(Contd.)

Table 6.1 *(Contd.)*

Sl. No.	Themes	Techniques
338.	Levels of livestock development	Choropleth
339.	Levels of development	Choropleth

6.2.4 Level of Data and Periodicity of Revision

The suggested level of data is block and the periodicity of revision is 5 years.

6.2.5 Explanatory Note to Maps

It may be recalled that 52 per cent of the users want explanatory note to accompany maps. It is proposed that, each map ought to have an explanatory note about the nature and computation of data, method of presentation required for comprehending the contents of the map and a brief description of the maps.

6.3 A MODEL RESOURCE ATLAS OF DINDIGUL DISTRICT

The map themes numbering 258, selected for District Level Planning include 49 themes of the maps, which were circulated among the map users at district level, as explained in Chapter 5. On the basis of the interaction with these map users, the researcher revised these maps in 34 plates as a model set of resource maps for application in district planning.

The base map of this district was prepared based on the maps (58 F and 58 J) in 1:250000 scale published by Survey of India in 1976. They also provided the base details like roads, railways, *taluk* boundaries, location of settlements, streams etc., and they have been used for drawing relief, slope and drainage maps. The paper size of the original map was 50cm X 40cm and has been reduced to 17.5cm X 10cm so as to place them in the book.

All these maps have been drawn based on the data collected from the departments functioning in the district. Maps on population were drawn based on Census of India, 1981 and others on 1989-90 data. The data at block level was decided mainly on the basis of availability of data and requirements of the user. With regard to techniques of mapping, in all choropleth plates, number of classes have been fixed at four.

6.3.1 Profile of Dindigul District

(a) **Location**: Dindigul District was bifurcated from old Madurai District on 15th September, 1985 with Dindigul town as its headquarters – recently renamed as Dindigul District. This district is situated between 77°30' E to 78° 20' E and 10° 02' N to 10° 50' N. This district consists of two Revenue divisions comprising 14 blocks.

(b) **Climate**: In the plains, semi-arid tropical monsoon type of climate is prevailing. However, hilly region records low temperature and fairly heavy rainfall. North-East monsoon is the main source of water in the district.

(c) **Population**: Total population of the district is 17,60,601 (1991) of which more than 78 per cent of them live in rural areas. Density of population is 417 per sq. km. Of the total population Scheduled Castes population accounts for over 20 per cent. About 50 per cent of the population is literate and the sex-ratio is 977. Occupational structure of the district shows that, 61 per cent are engaged in primary, 29 per cent in secondary activities and the rest in service sector.

(d) **Land use and cropping pattern**: Total geographical area is 6480.18 sq.km. Of the total area, 56.6 per cent is net sown area and about 19 per cent is forest. Only 9 per cent of the net sown area is cultivated more than once. Paddy is the principal crop (37 per cent) followed by oilseeds (23 per cent) and fruits and vegetables (24 per cent). In the plain region the important crops are paddy, cholam, cumbu, millets, oilseeds, cotton, sugarcane, onion etc. In Kodaikanal and Sirumalai hills, the major crops are coffee, citrus fruits etc.

(e) **Irrigation**: In this district the predominant source of irrigation is well. The important irrigation schemes are Palar, Porondalar, Kodaganar and Manjalar. In the plains, dry farming is practiced with the exception of Shanmuganadhi and Kodaganar river basins.

(f) **Industry** : Some of the fast developing industrial activities of recent origin are lime kiln, brick kiln, spinning and paper mills. Other traditional but important industries are leather tanning, lock making, brassware industry, tobacco and cigar making, jewellery work, dyeing, art silk weaving etc.

(g) **Infrastructure facilities**: Dindigul and Palani are the two important educational centres with arts and science colleges, professional colleges and polytechnics. The district has a University, *viz.* Gandhigram Rural Institute (Deemed University) at Gandhigram.

There are 1,351 schools functioning in the district. The district houses nine hospitals, 31 primary health centres and 307 sub-centres. Agricultural produces are marketed in six permanent and six regulated markets along with seven unregulated markets and seven shandies. With regard to veterinery facilities, four hospitals, 27 dispensaries, 11 key village sub-centres and one poultry extension centre are providing services to the livestock population.

The district has a good road network, well connected with other parts of the State. National Highway No. 7 and 47 run through the district. Surfaced road length is 3,071 km while unsurfaced length is 3,071.37 km. Metre guage railway line connects the district with Coimbatore, Trichy and Madurai while broad guage links with Karur and Madurai.

The Maps of this model resource atlas deal with themes in natural resources, ground water/irrigation, population, landuse/agriculture, industries and infrastructural facilities. They are as follows :

Map No.	**Theme**
6.1.	Location of Dindigul District
I.	**Natural Resources**
6.2.	Relief
6.3.	Slope
6.4.	Soils
6.5.	Average Annual Temperature
6.6.	Annual Rainfall
6.7.	Seasonal Rainfall
6.8.	Drainage
II.	**Ground Water/Irrigation**
6.9.	Depth of Ground Water Table
6.10.	Ground Water Quality
6.11.	Extent of Irrigation
	Well Irrigation
	Tank Irrigation
	Other Sources of Irrigation
III.	**Population**
6.12.	Density of Population
6.13.	Occupational Structure
6.14.	Literates

6.15. Birth, Death and Infant Mortality Rates

IV. Agriculture

6.16. Land use

6.17. Barren and Uncultivable Land (as percentage of total geog. area)
Area under Cultivable Waste (as percentage of total geog. area)
Area under Current Fallows (as percentage of total geog. area)
Other Fallow Lands (as percentage of total geog. area)

6.18. Permanent Pastures and other Grazing Lands (as percentage of total geog. area)
Land under Non-Agricultural Uses (as percentage of total geog. area)
Area under Miscellaneous Tree Crops (as percentage of total geog. area)
Net Sown Area (as percentage of total geog. area)

6.19. Forest

6.20. Area under Important Crops : Bajra
Sorghun
Paddy
Groundnut

6.21. Man-Land Ratio
Intensity of Cropping

6.22. Landholdings : Area
Number

6.23. Agricultural Machinery and Implements

6.24. Infrastructural Facilities for Agriculture

6.25. Markets

V. Industries

6.26. Industries

6.27. Agro-based Industries

6.28. Artisan-based Industries

VI. Infrastructural Facilities

6.29. Financial Institutions

6.30. Cooperative Societies

6.31. Centres of Education

6.32. Health Facilities

6.33. Veterinary Facilities

6.34. Roads and Railways

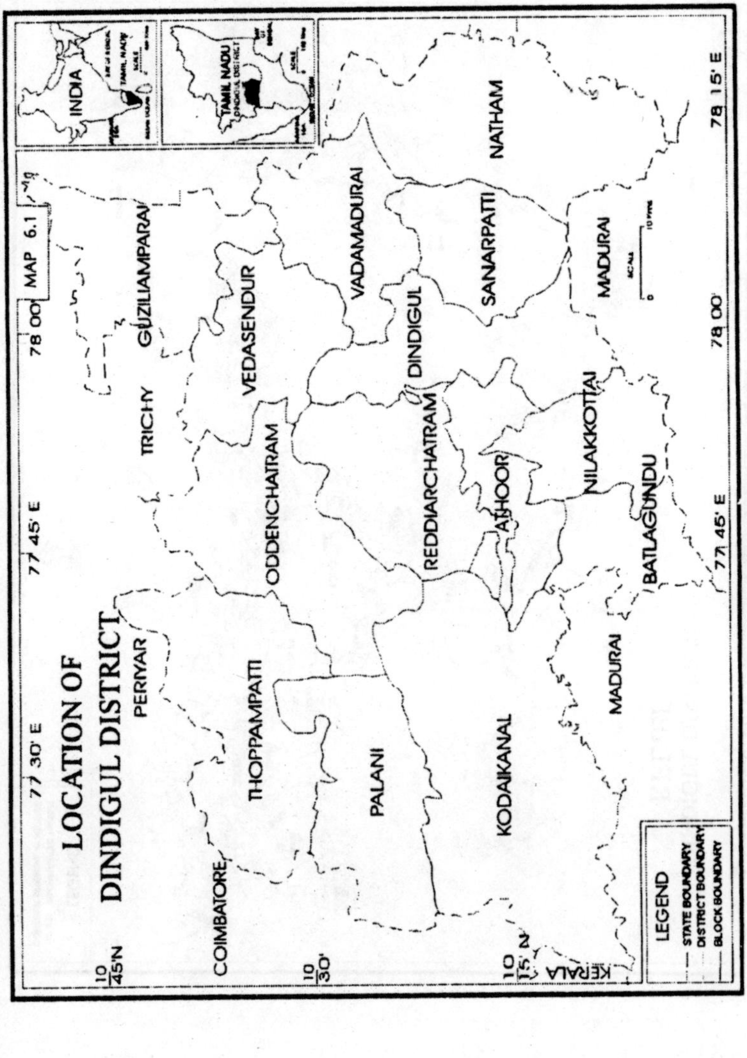
LOCATION OF
DINDIGUL DISTRICT
MAP 6.1
INDIA
TAMIL NADU
77 30' E
77 45' E
78 00'
78 15' E
10 45'N
10 30'
10 15' N
COIMBATORE
PERIYAR
TRICHY
GUZILIAMPARAI
VEDASENDUR
ODDENCHATRAM
THOPPAMPATTI
PALANI
KODAIKANAL
REDDIARCHATRAM
DINDIGUL
VADAMADURAI
SANARPATTI
NATHAM
ATHOOR
NILAKKOTTAI
BATLAGUNDU
MADURAI
MADURAI
KERALA
LEGEND
STATE BOUNDARY
DISTRICT BOUNDARY
BLOCK BOUNDARY

Map 6.1

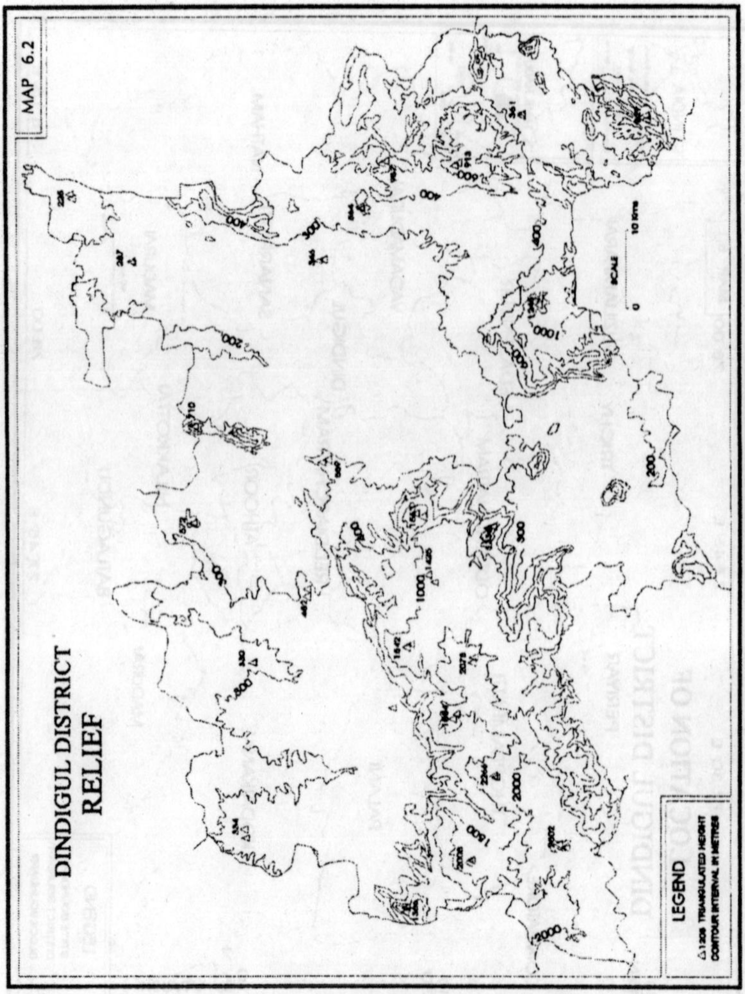
DINDIGUL DISTRICT
RELIEF
MAP 6.2
LEGEND
TRIANGULATED HEIGHT
CONTOUR INTERVAL IN METRES

Map 6.2

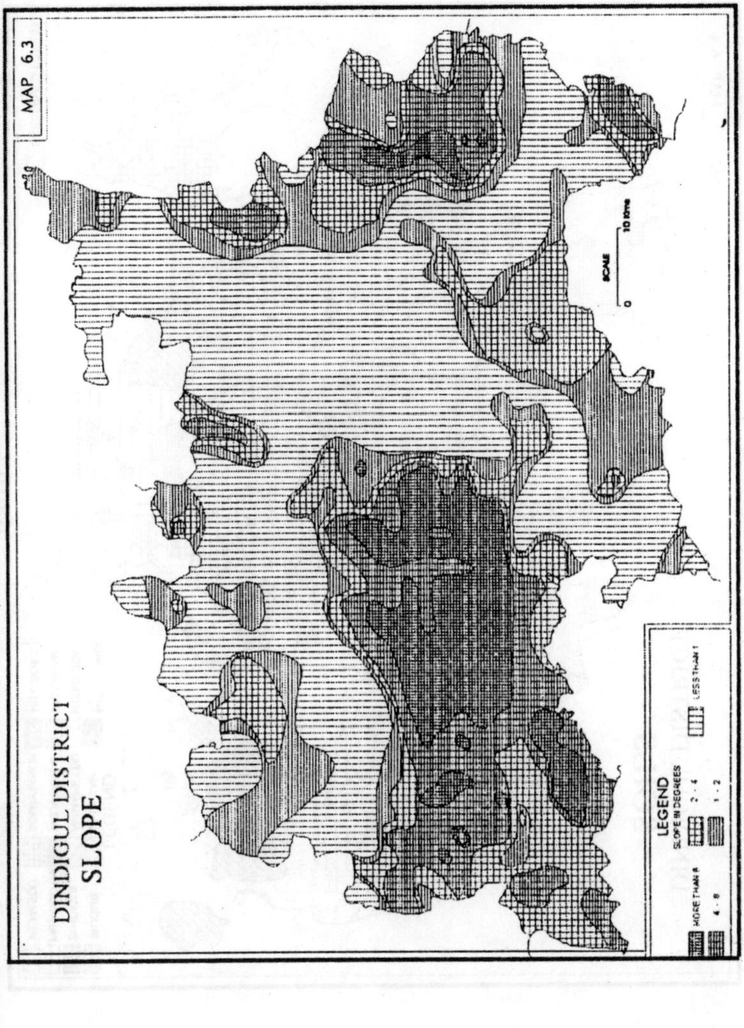
MAP 6.3
DINDIGUL DISTRICT
SLOPE
LEGEND
SLOPE IN DEGREES
MORE THAN 8
4 - 8
2 - 4
1 - 2
LESS THAN 1
SCALE
0
10 kms

Map 6.3

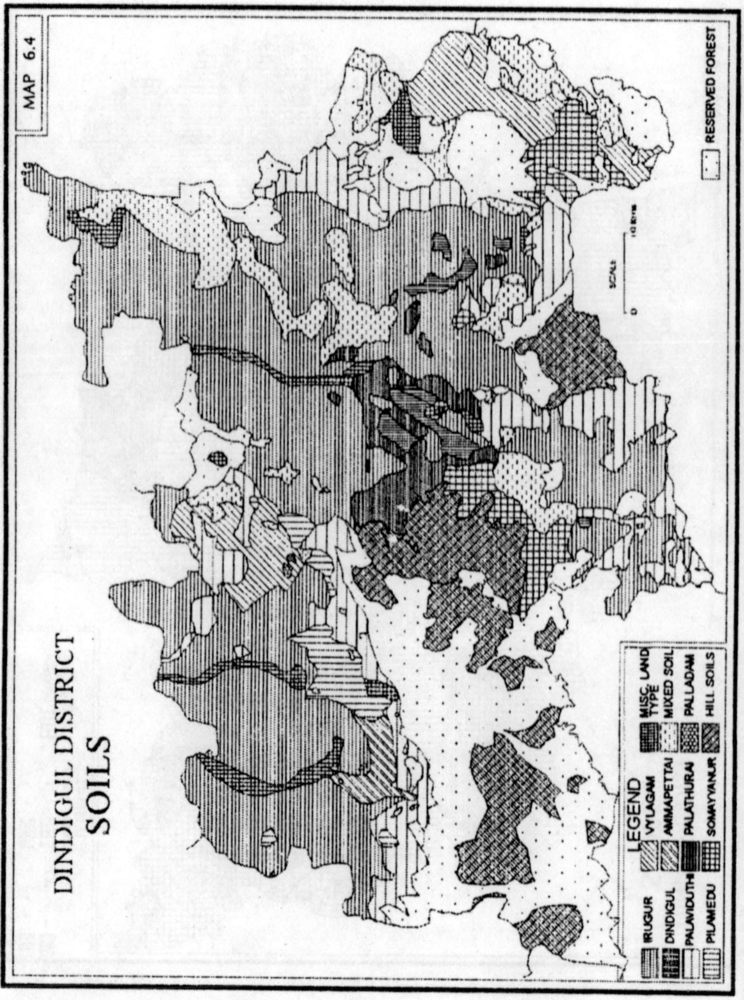
MAP 6.4
DINDIGUL DISTRICT
SOILS
LEGEND
IRUGUR
DINDIGUL
PALAVIDUTHI
PILAMEDU
VYLAGAM
AMMAPETTAI
PALATHURAI
SOMAYYANUR
MISC. LAND TYPE
MIXED SOIL
PALLADAM
HILL SOILS
RESERVED FOREST
SCALE

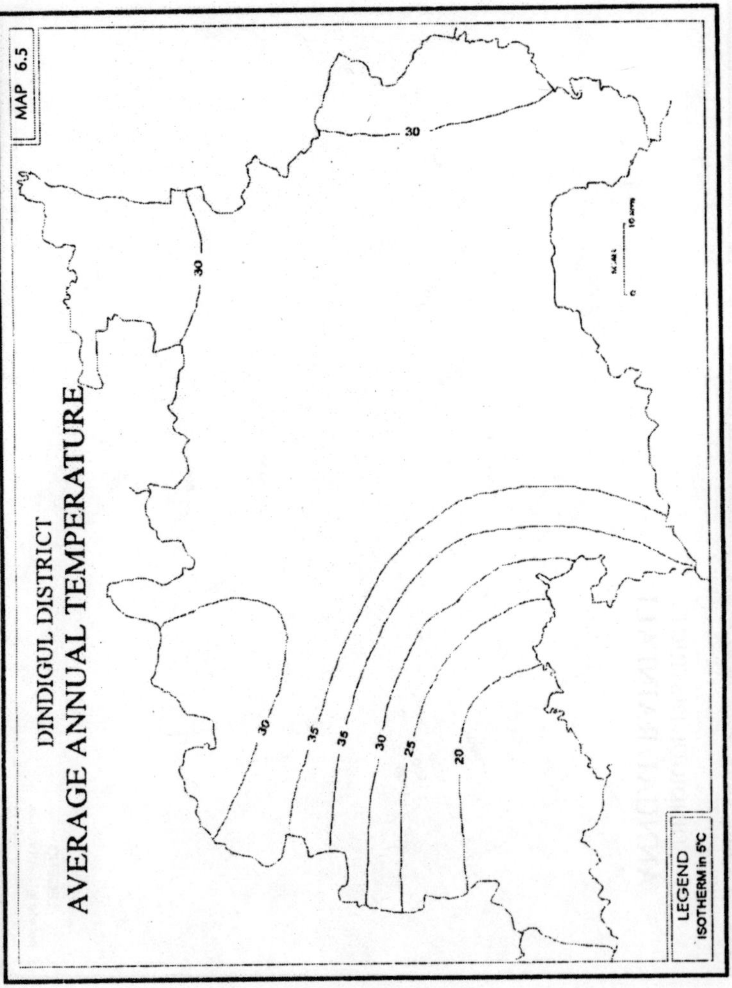
MAP 6.5
DINDIGUL DISTRICT
AVERAGE ANNUAL TEMPERATURE
30
30
30
35
35
30
25
20
LEGEND
ISOTHERM in 5°C

Map 6.5

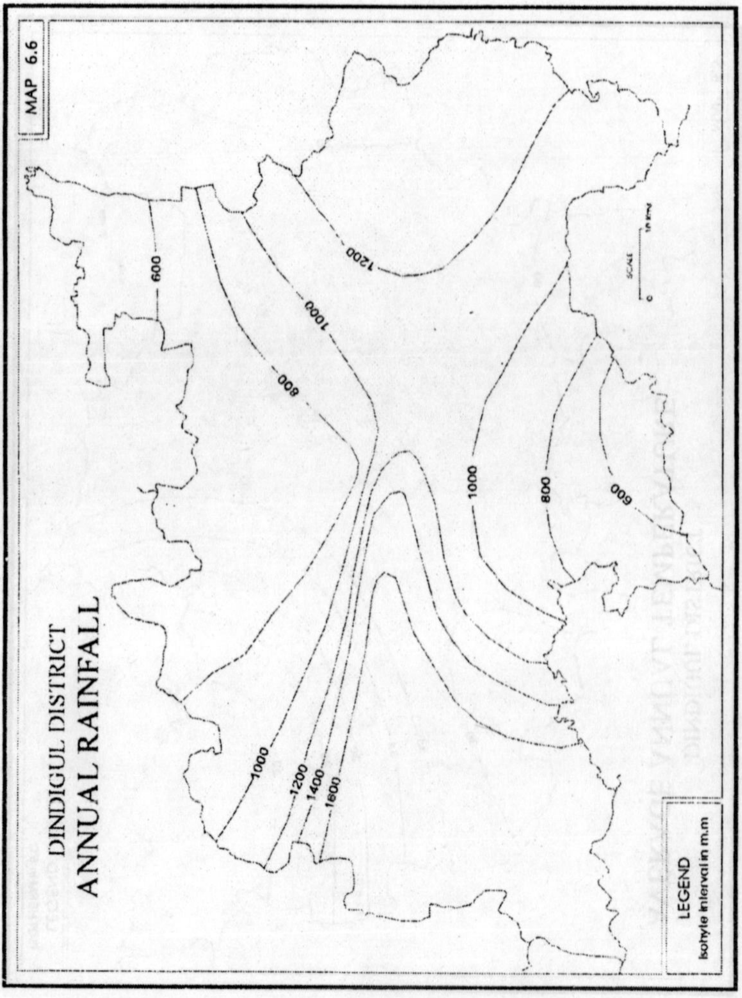
MAP 6.6
DINDIGUL DISTRICT
ANNUAL RAINFALL
600
800
1000
1200
1000
800
600
1000
1200
1400
1600
LEGEND
Isohyte interval in m.m
SCALE

Map 6.6

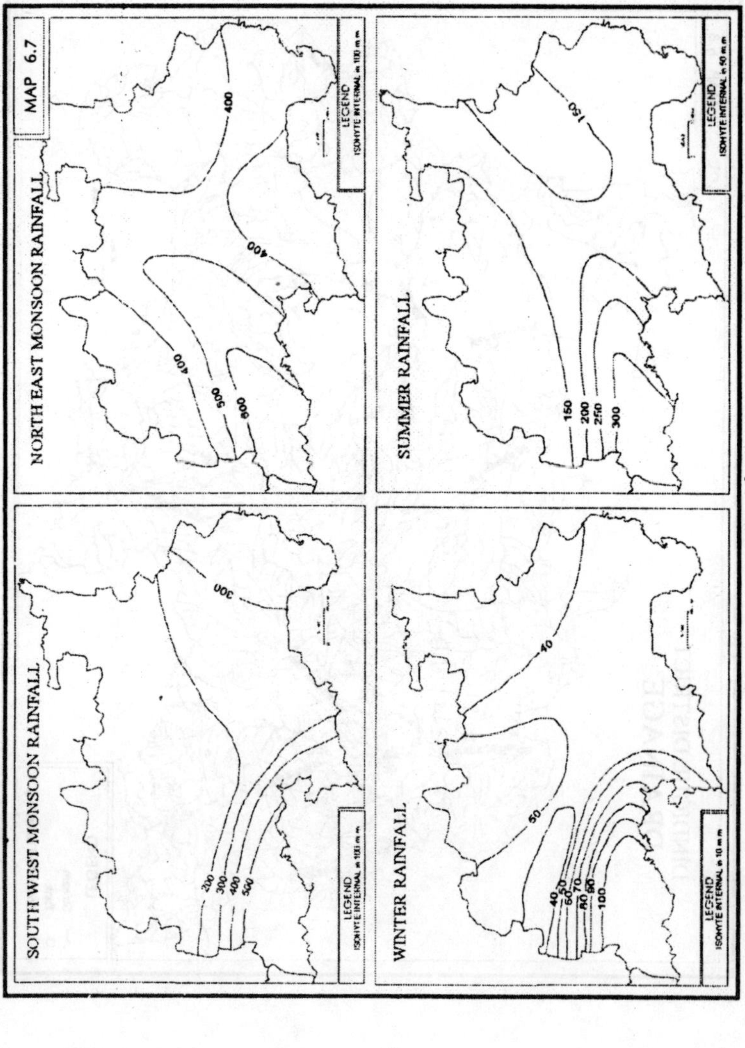
MAP 6.7
NORTH EAST MONSOON RAINFALL
400
400
400
500
600
LEGEND
ISOHYTE INTERNAL in 100 m m
SOUTH WEST MONSOON RAINFALL
300
200
300
400
500
LEGEND
ISOHYTE INTERNAL in 100 m m
SUMMER RAINFALL
150
150
200
250
300
LEGEND
ISOHYTE INTERNAL in 50 m m
WINTER RAINFALL
40
50
40
50
60
70
80
90
100
LEGEND
ISOHYTE INTERNAL in 10 m m

Map 6.7

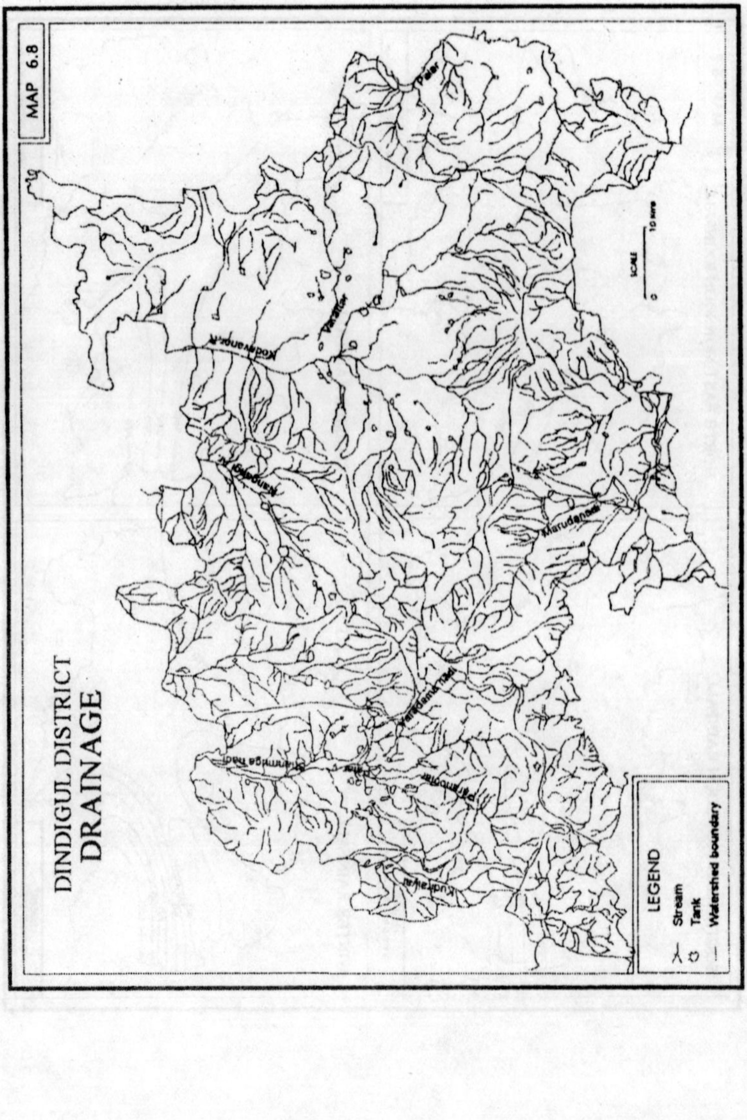
MAP 6.8
DINDIGUL DISTRICT
DRAINAGE
LEGEND
Stream
Tank
Watershed boundary
SCALE
Palar
Marudanadi

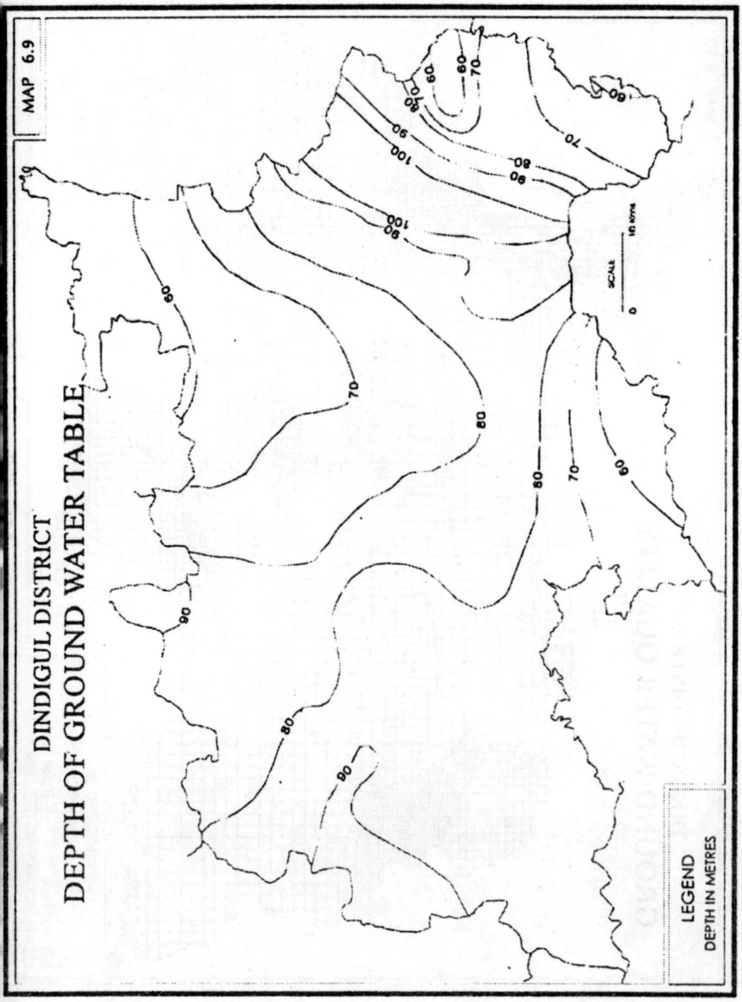
MAP 6.9
DINDIGUL DISTRICT
DEPTH OF GROUND WATER TABLE
LEGEND
DEPTH IN METRES
SCALE
60
70
80
90
100

Map 6.9

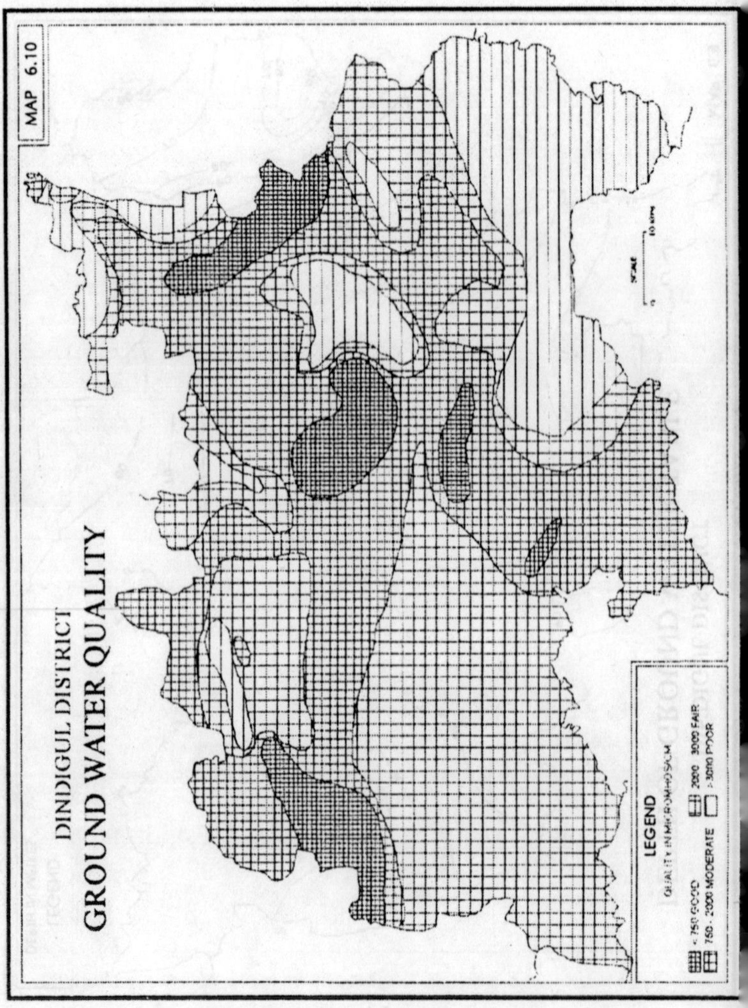
MAP 6.10
DINDIGUL DISTRICT
GROUND WATER QUALITY
LEGEND
SCALE

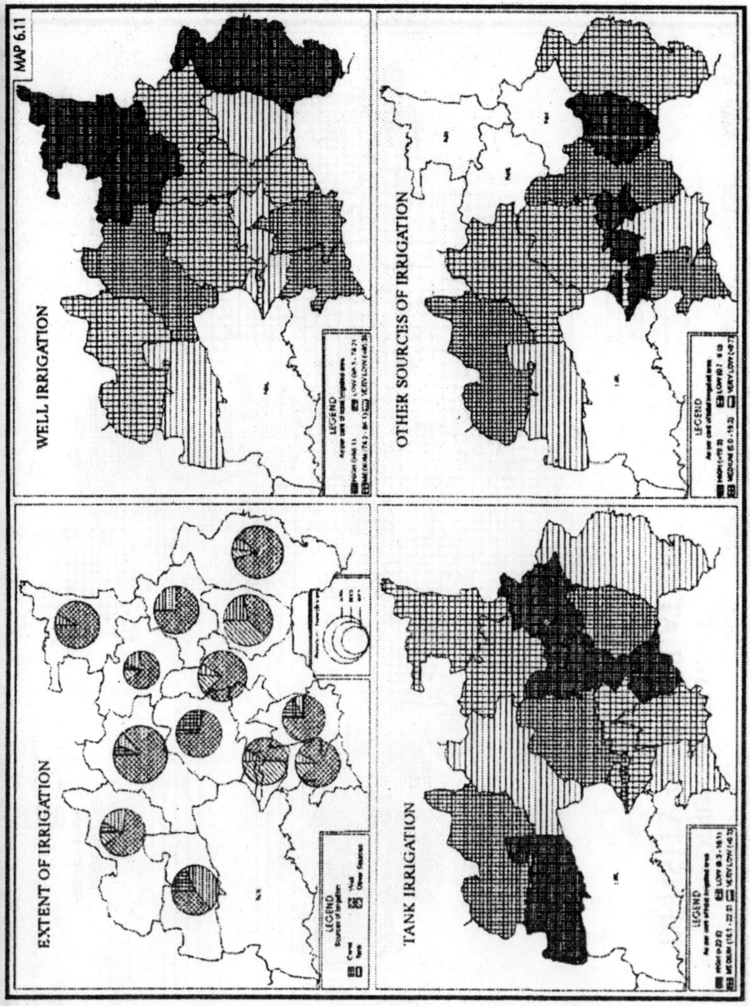
MAP 6.11
EXTENT OF IRRIGATION
WELL IRRIGATION
TANK IRRIGATION
OTHER SOURCES OF IRRIGATION
LEGEND

Map 6.11

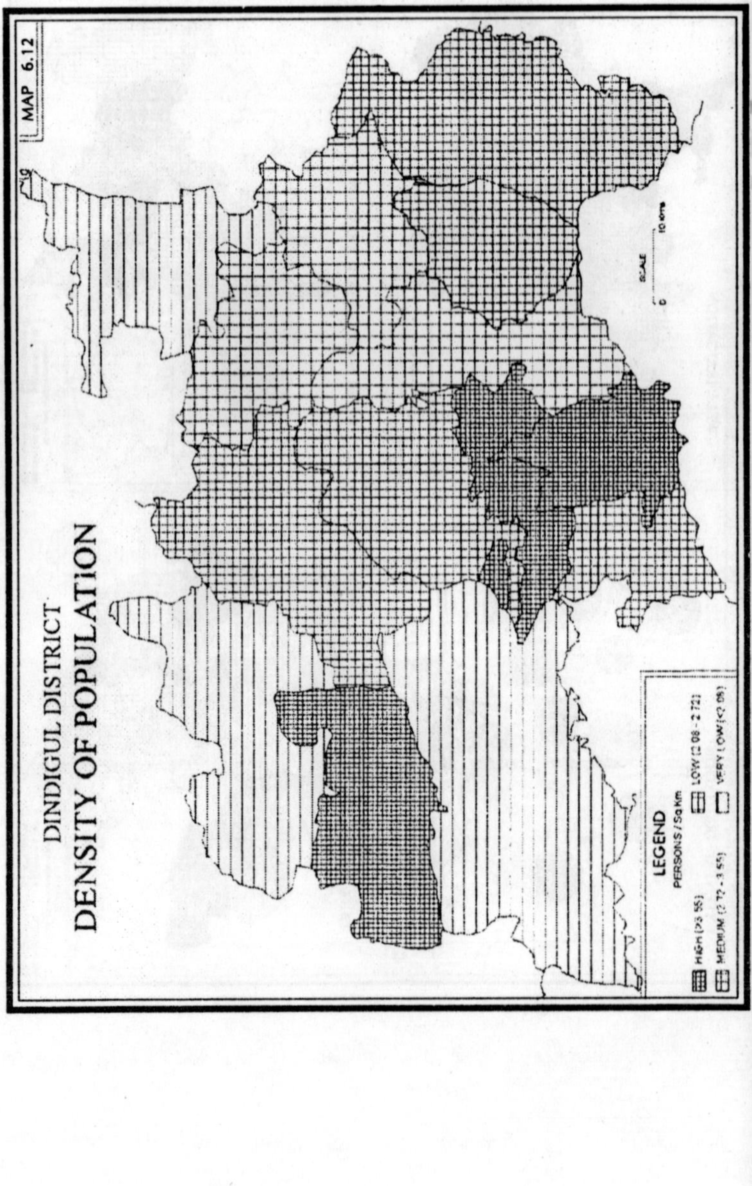
MAP 6.12
DINDIGUL DISTRICT
DENSITY OF POPULATION
LEGEND
PERSONS / Sq Km

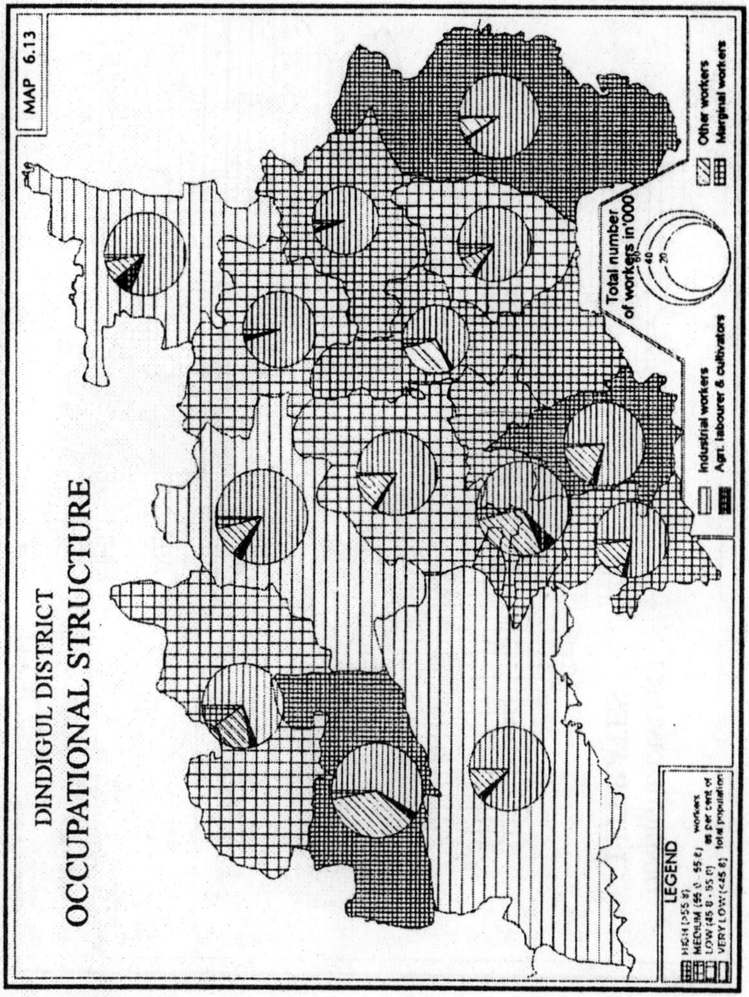
DINDIGUL DISTRICT
OCCUPATIONAL STRUCTURE
MAP 6.13
LEGEND
workers as per cent of total population
Industrial workers
Agri. labourer & cultivators
Other workers
Marginal workers
Total number of workers in'000
60
40
20

Map 6.13

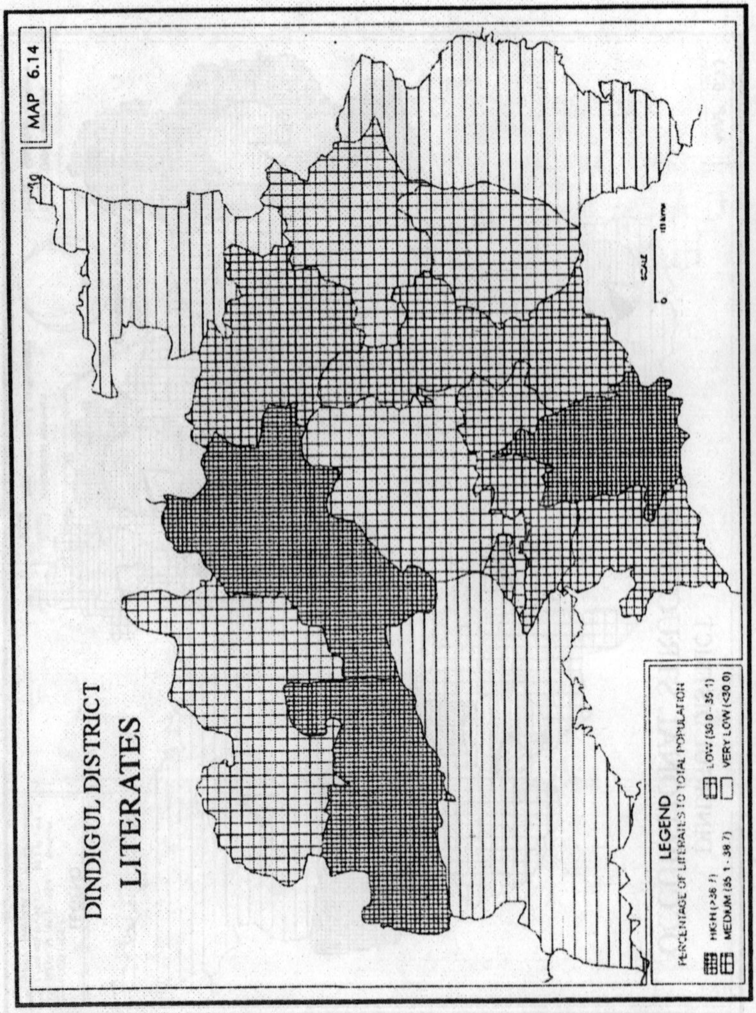
MAP 6.14
DINDIGUL DISTRICT
LITERATES
SCALE
LEGEND
PERCENTAGE OF LITERATES TO TOTAL POPULATION
HIGH (>38.7)
MEDIUM (35.1 - 38.7)
LOW (30.0 - 35.1)
VERY LOW (<30.0)

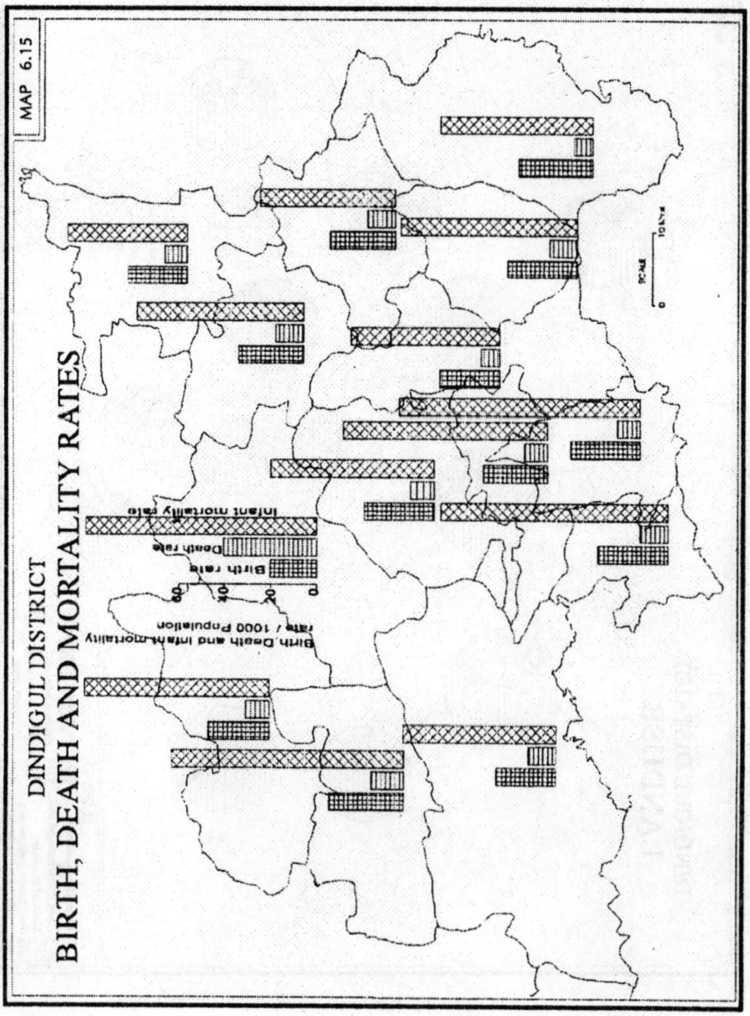
MAP 6.15
DINDIGUL DISTRICT
BIRTH, DEATH AND MORTALITY RATES
Birth, Death and Infant mortality rate / 1000 Population
60
40
20
0
Birth rate
Death rate
Infant mortality rate

Map 6.15

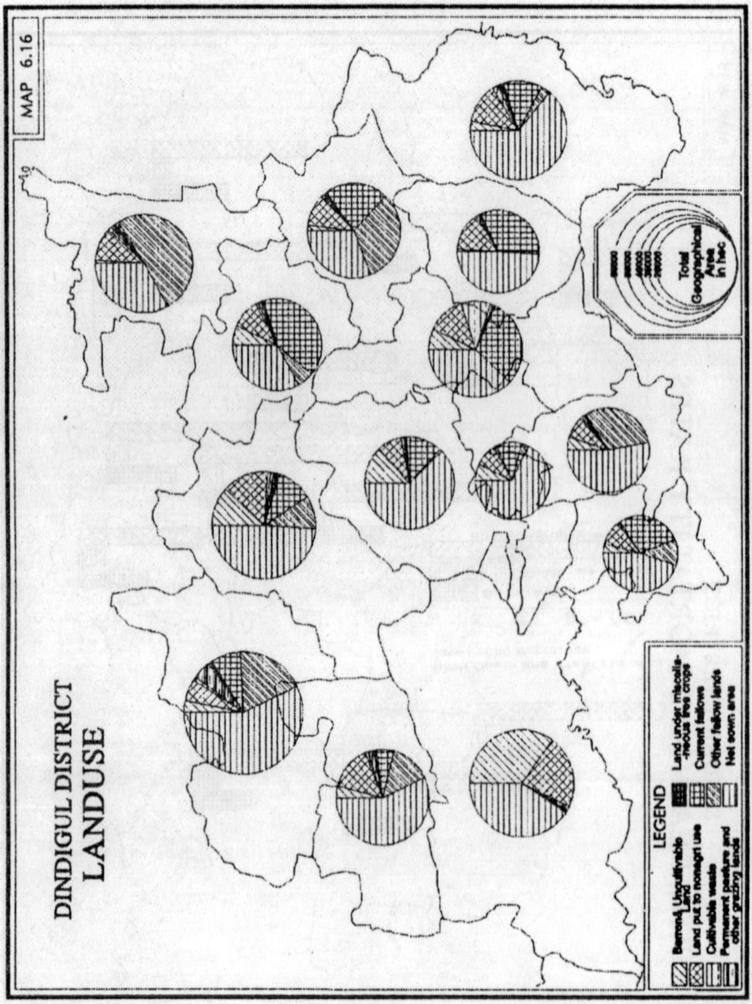
MAP 6.16
DINDIGUL DISTRICT
LANDUSE
Total Geographical Area in hac
LEGEND
Barren& Uncultivable Land
Land put to nonagri use
Cultivable waste
Permanent pasture and other grazing lands
Land under miscella-neous tree crops
Current fallows
Other fallow lands
Net sown area

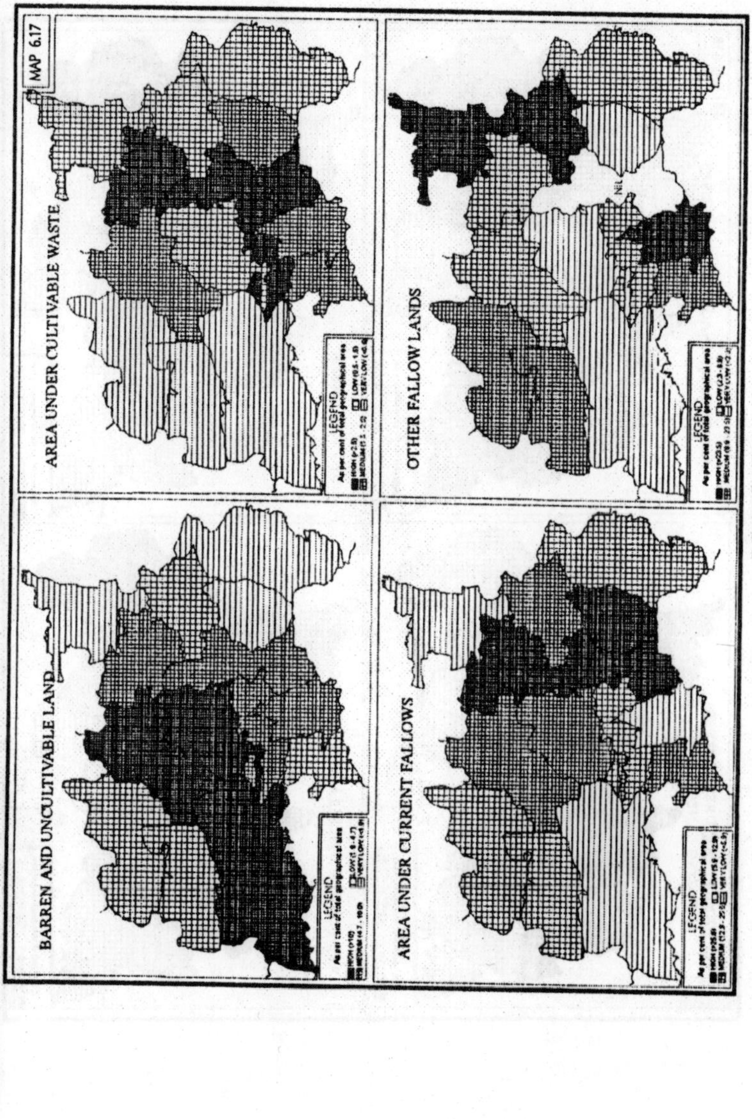
MAP 6.17
AREA UNDER CULTIVABLE WASTE
BARREN AND UNCULTIVABLE LAND
OTHER FALLOW LANDS
AREA UNDER CURRENT FALLOWS
NIL
LEGEND
LEGEND
LEGEND
LEGEND

Map 6.17

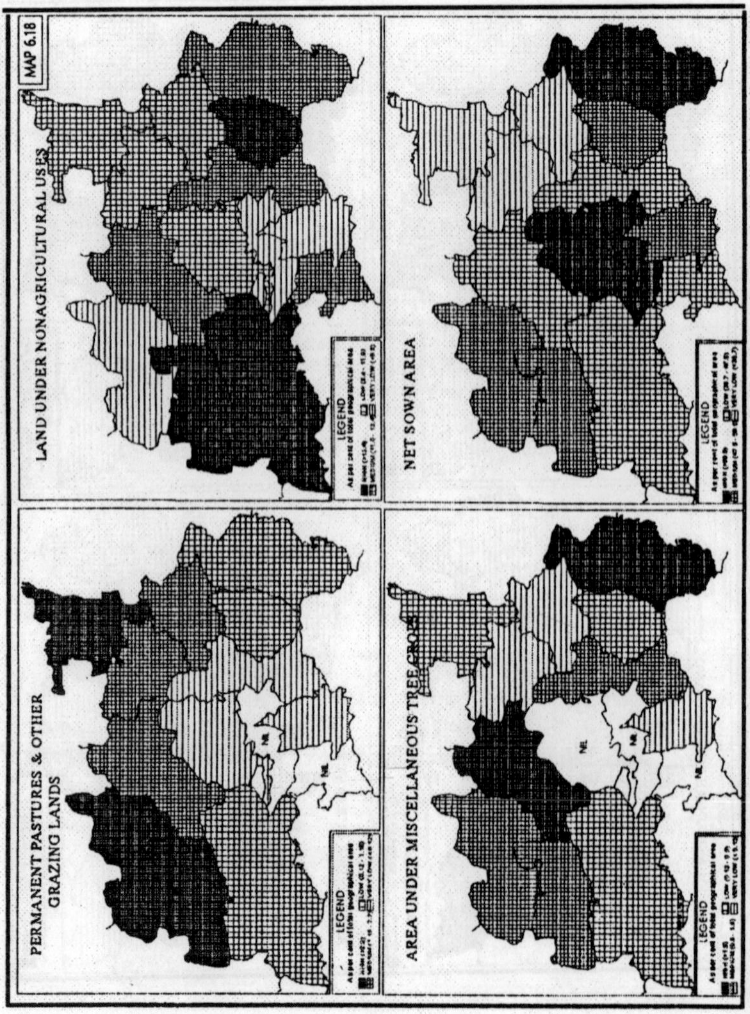
MAP 6.18
LAND UNDER NONAGRICULTURAL USES
LEGEND
NET SOWN AREA
LEGEND
PERMANENT PASTURES & OTHER GRAZING LANDS
NIL
NIL
LEGEND
AREA UNDER MISCELLANEOUS TREE CROPS
NIL
NIL
NIL
LEGEND

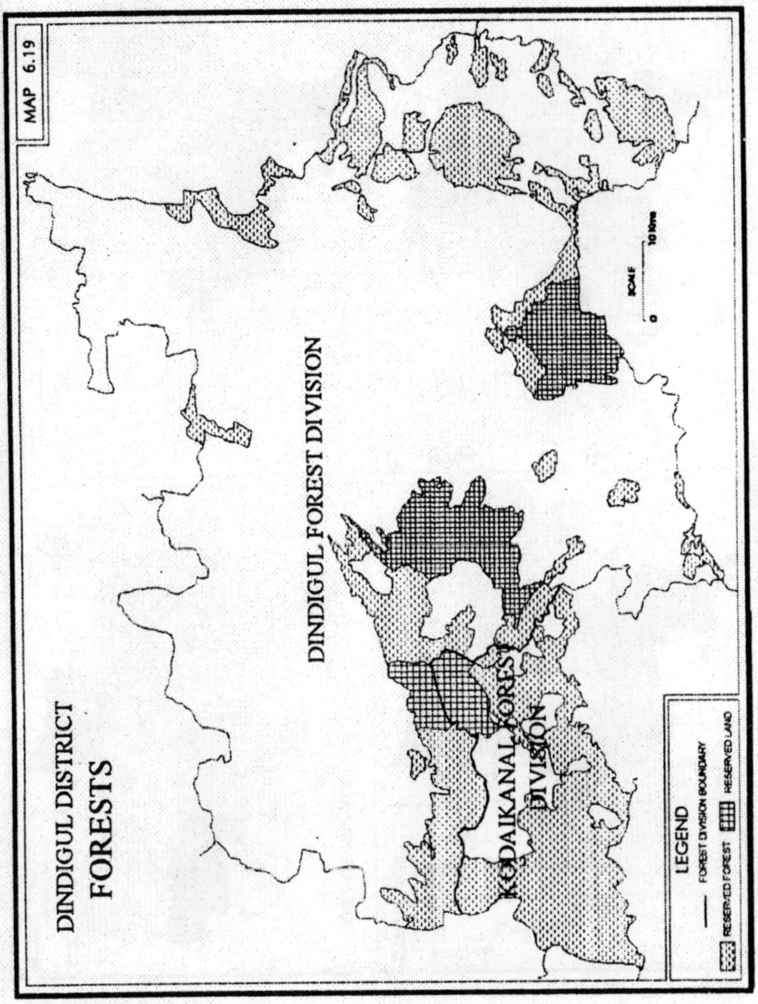
MAP 6.19
DINDIGUL DISTRICT
FORESTS
DINDIGUL FOREST DIVISION
KODAIKANAL FOREST DIVISION
LEGEND
FOREST DIVISION BOUNDARY
RESERVED FOREST
RESERVED LAND

Map 6.19

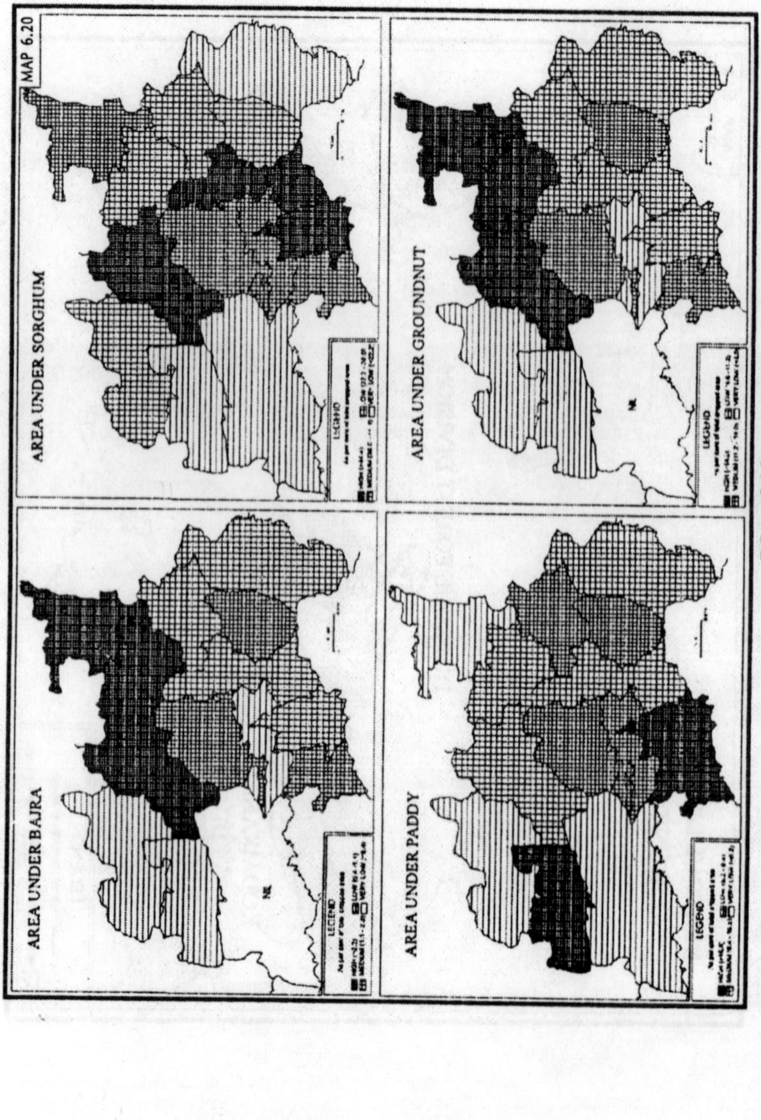
MAP 6.20
AREA UNDER SORGHUM
AREA UNDER BAJRA
AREA UNDER GROUNDNUT
AREA UNDER PADDY
LEGEND

Map 6.20

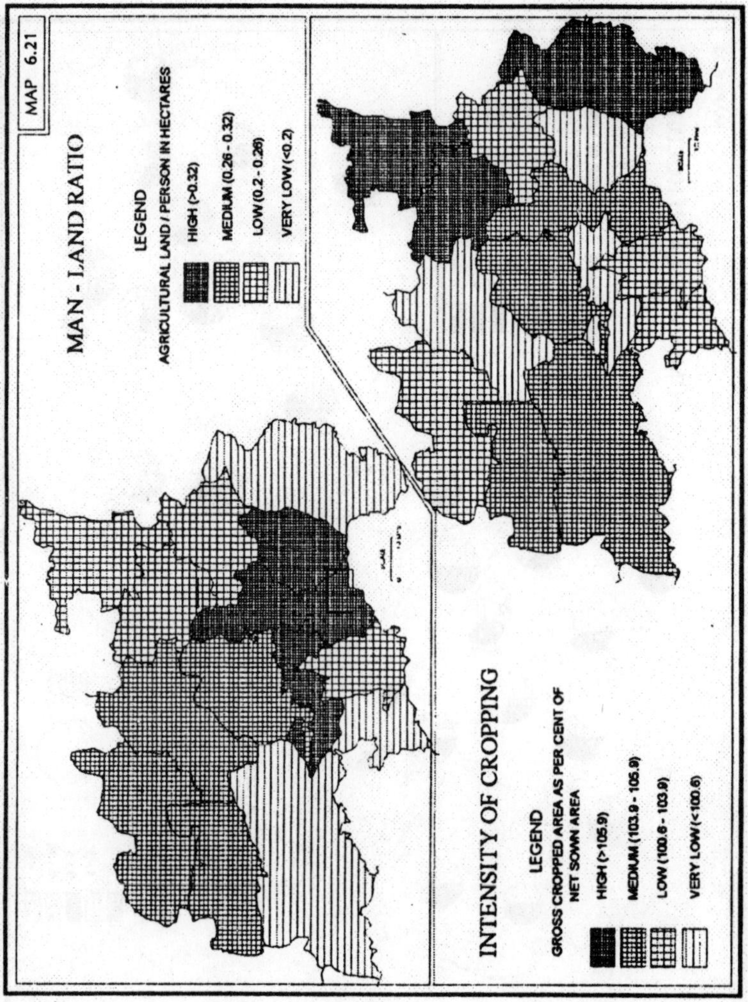
MAP 6.21
MAN - LAND RATIO
LEGEND
AGRICULTURAL LAND / PERSON IN HECTARES
HIGH (>0.32)
MEDIUM (0.26 - 0.32)
LOW (0.2 - 0.26)
VERY LOW (<0.2)
INTENSITY OF CROPPING
LEGEND
GROSS CROPPED AREA AS PER CENT OF NET SOWN AREA
HIGH (>105.9)
MEDIUM (103.9 - 105.9)
LOW (100.6 - 103.9)
VERY LOW (<100.6)

Map 6.21

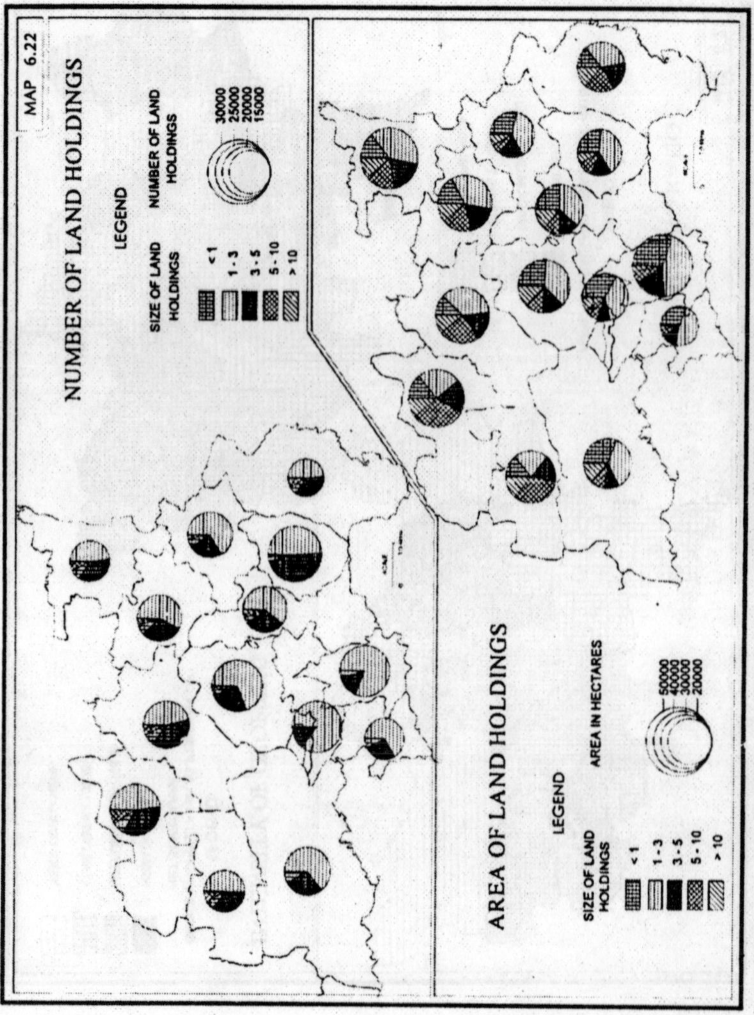
MAP 6.22
NUMBER OF LAND HOLDINGS
LEGEND
SIZE OF LAND HOLDINGS
<1
1-3
3-5
5-10
>10
NUMBER OF LAND HOLDINGS
30000
25000
20000
15000
AREA OF LAND HOLDINGS
LEGEND
SIZE OF LAND HOLDINGS
<1
1-3
3-5
5-10
>10
AREA IN HECTARES
50000
40000
30000
20000

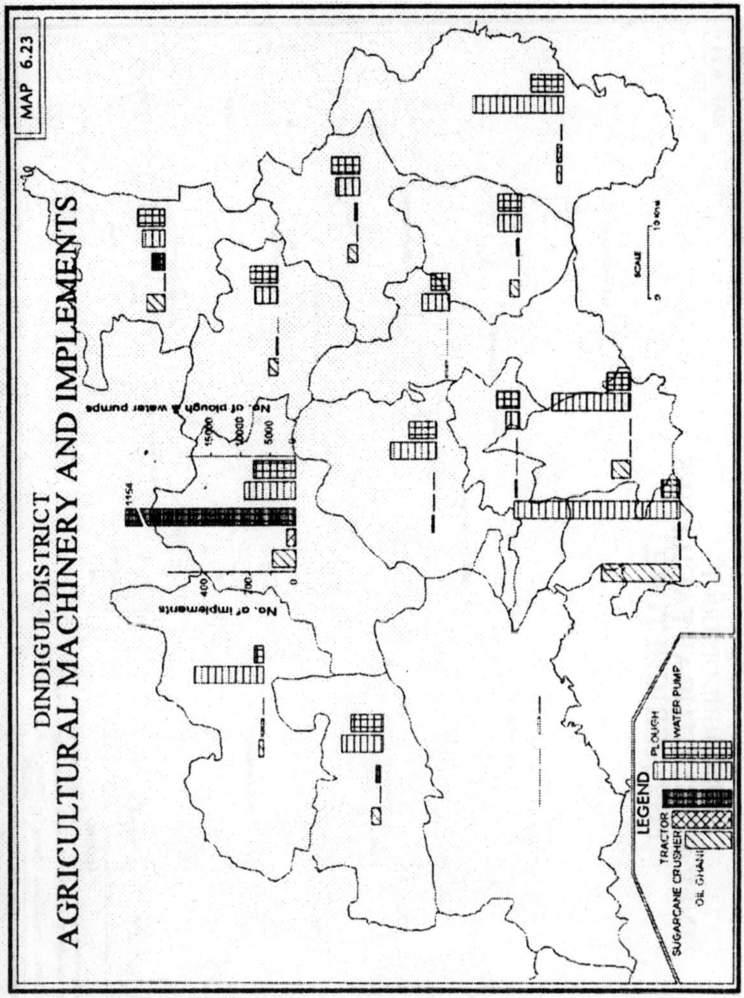
MAP 6.23
DINDIGUL DISTRICT
AGRICULTURAL MACHINERY AND IMPLEMENTS
No. of plough & water pumps
15000
10000
5000
0
1154
No. of implements
400
200
0
SCALE
LEGEND
TRACTOR
SUGARCANE CRUSHER
OIL GHANI
PLOUGH
WATER PUMP

Map 6.23

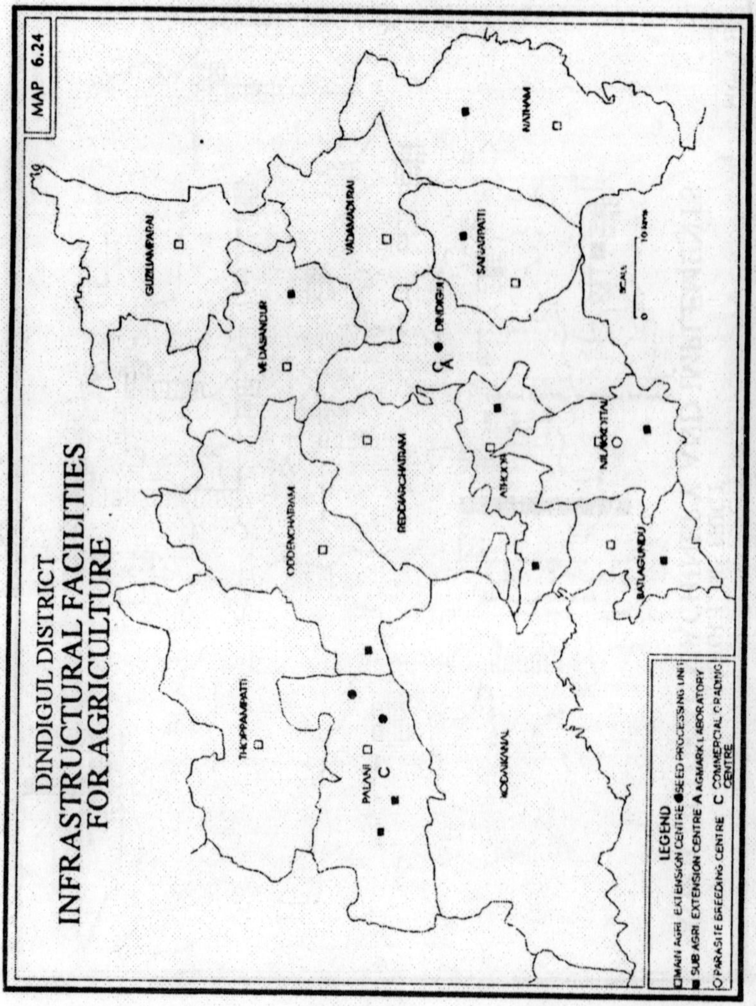
MAP 6.24
DINDIGUL DISTRICT
INFRASTRUCTURAL FACILITIES
FOR AGRICULTURE
GUZILIAMPARAI
VADAMADURAI
NATHAM
VEDASANDUR
SANARPATTI
DINDIGUL
ODDENCHATRAM
REDDIARCHATRAM
ATHOOR
NILAKKOTTAI
BATLAGUNDU
THOPPAMPATTI
PALANI
KODAIKANAL
SCALE
0
10 kms
LEGEND
MAIN AGRI. EXTENSION CENTRE
SEED PROCESSING UNIT
SUB AGRI. EXTENSION CENTRE
A AGMARK LABORATORY
PARASITE BREEDING CENTRE
C COMMERCIAL GRADING CENTRE

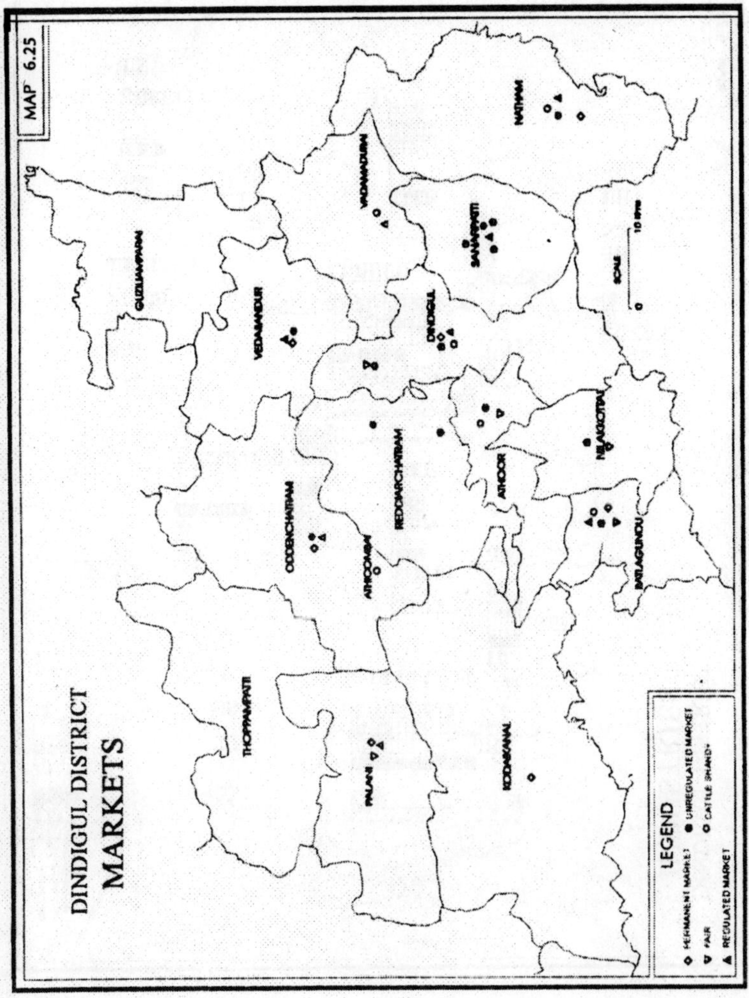

Map 6.25

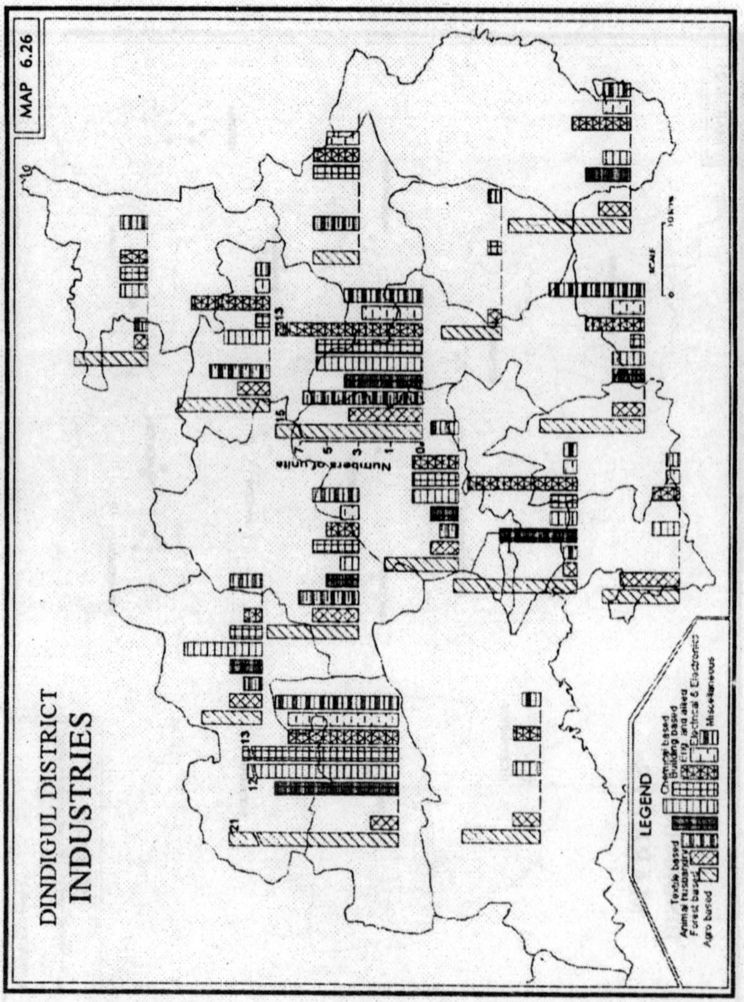
MAP 6.26
DINDIGUL DISTRICT
INDUSTRIES
Numbers of Units
LEGEND
Textile based
Animal husbandry
Forest based
Agro based
Chemical based
Building based
Engg. and allied
Electrical & Electronics
Miscellaneous
SCALE

Map 6.26

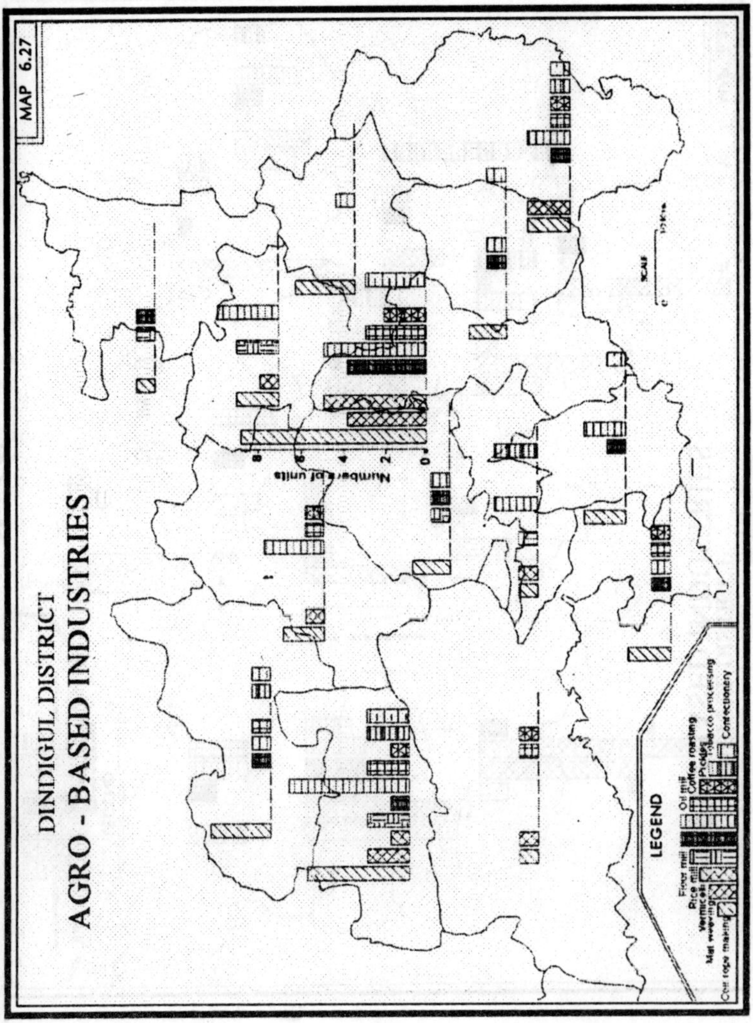
MAP 6.27
DINDIGUL DISTRICT
AGRO - BASED INDUSTRIES
Number of units
LEGEND
Flour mill
Rice mill
Vermicelli
Mat weaving
Coir rope making
Oil mill
Coffee roasting
Pickles
Tobacco processing
Confectionery

Map 6.27

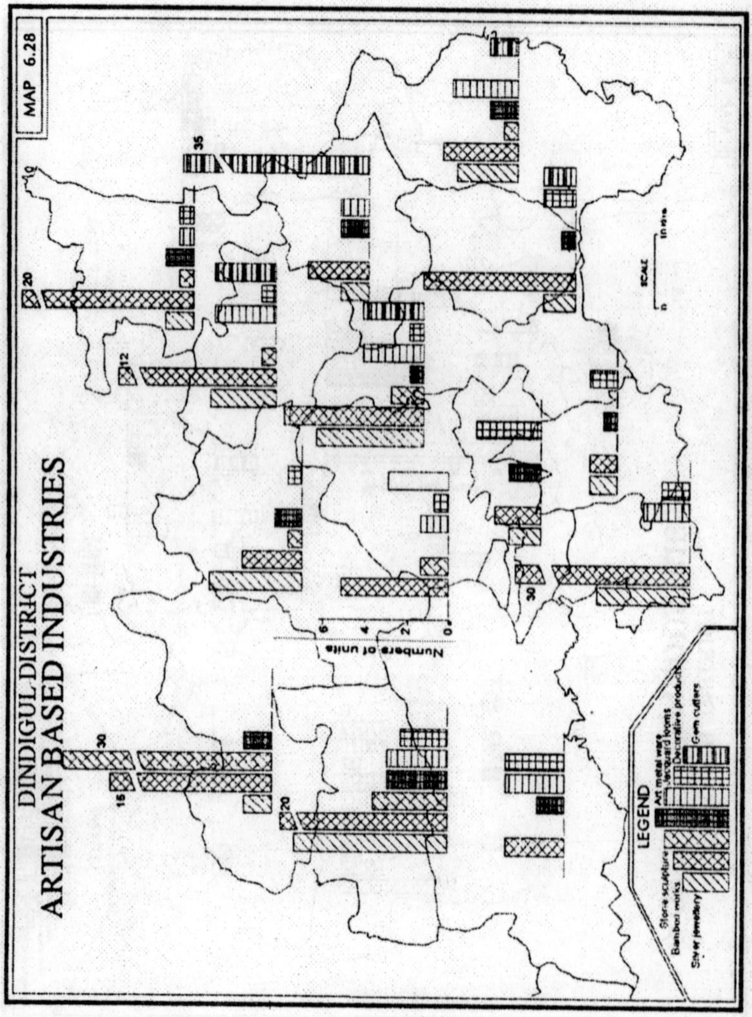
DINDIGUL DISTRICT
ARTISAN BASED INDUSTRIES
MAP 6.28
LEGEND
Stone sculpture
Bamboo works
Silver jewellery
Art metal ware
Jacquard looms
Decorative products
Gem cutters
Numbers of units
6
4
2
0
SCALE

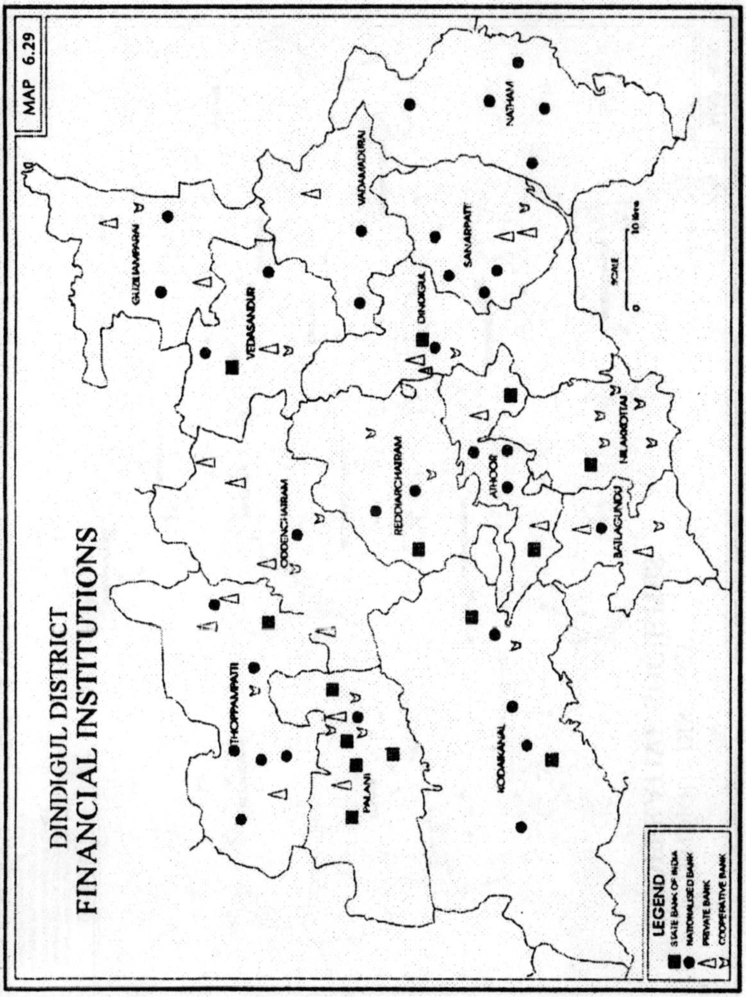
MAP 6.29
DINDIGUL DISTRICT
FINANCIAL INSTITUTIONS
GUZILIAMPARAI
VEDASANDUR
VADAMADURAI
NATHAM
SANARPATTI
DINDIGUL
REDDIARCHATRAM
ODDENCHATRAM
ATHOOR
NILAKKOTTAI
BATLAGUNDU
THOPPAMPATTI
PALANI
KODAIKANAL
SCALE
LEGEND
STATE BANK OF INDIA
NATIONALISED BANK
PRIVATE BANK
COOPERATIVE BANK

Map 6.29

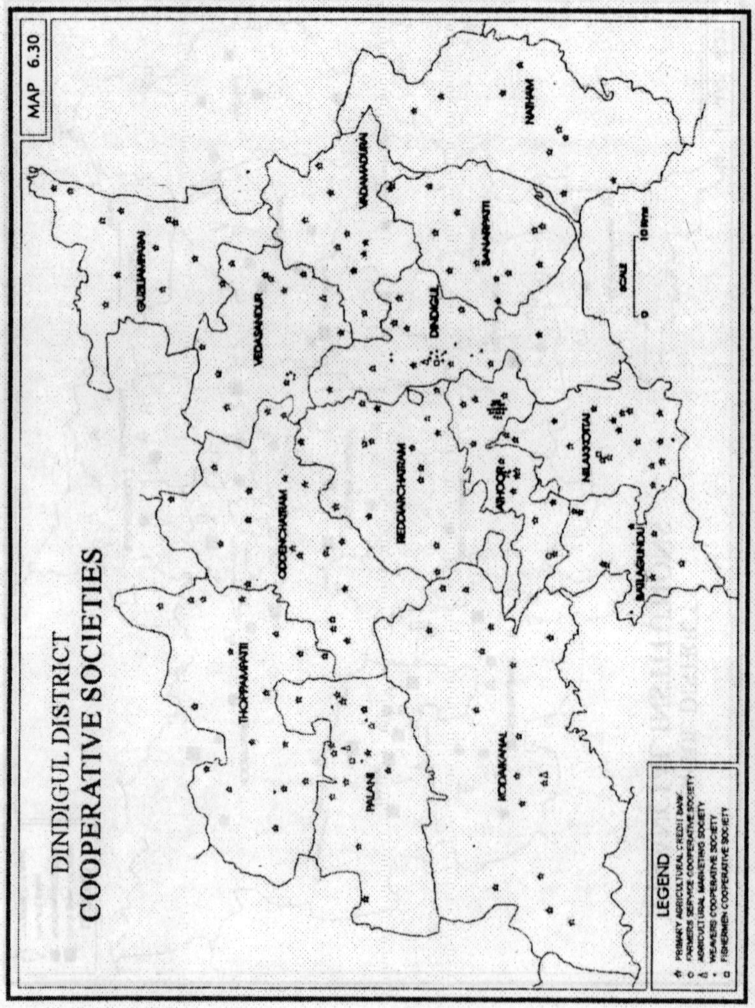
MAP 6.30
DINDIGUL DISTRICT
COOPERATIVE SOCIETIES
GUZILIAMPARAI
VADAMADURAI
NATHAM
SANARPATTI
DINDIGUL
VEDASANDUR
ODDENCHATRAM
REDDIARCHATRAM
ATHOOR
NILAKKOTTAI
BATLAGUNDU
THOPPAMPATTI
PALANI
KODAIKANAL
LEGEND

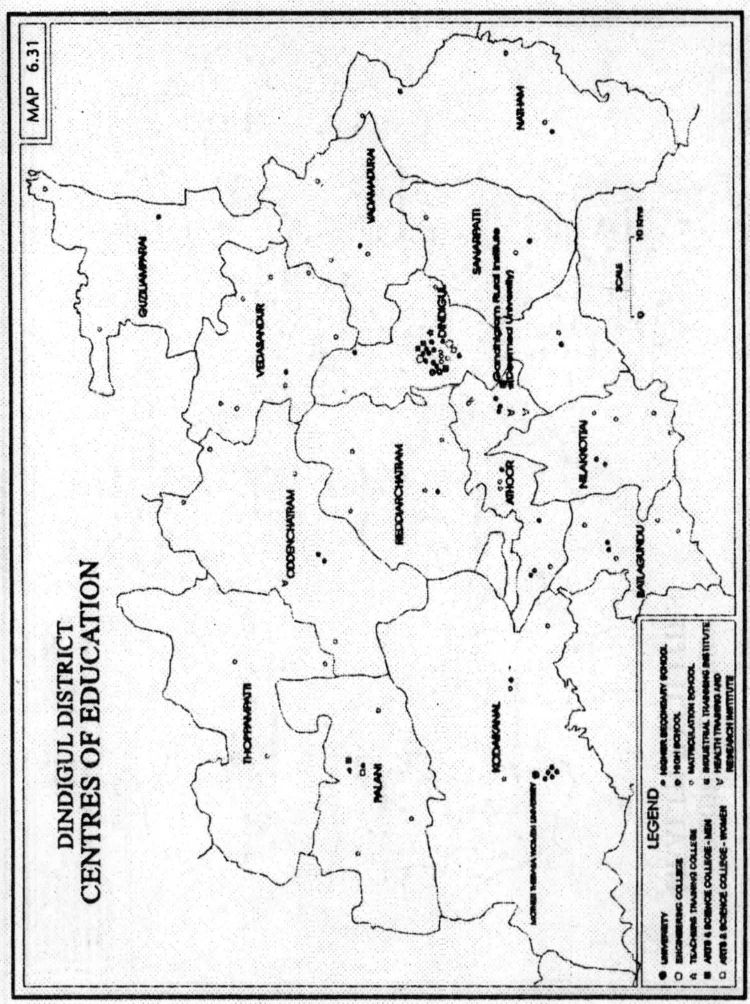
MAP 6.31
DINDIGUL DISTRICT
CENTRES OF EDUCATION
GUZILIAMPARAI
VEDASANDUR
VADAMADURAI
NATHAM
SANARPATI
DINDIGUL
Gandhigram Rural Institute (Deemed University)
ODDENCHATRAM
REDDIARCHATRAM
ATHOOR
NILAKKOTAI
BATLAGUNDU
PALANI
KODAIKANAL
SCALE
10 Kms
LEGEND
UNIVERSITY
ENGINEERING COLLEGE
TEACHING TRAINING COLLEGE
ARTS & SCIENCE COLLEGE - MEN
ARTS & SCIENCE COLLEGE - WOMEN
HIGHER SECONDARY SCHOOL
HIGH SCHOOL
MATRICULATION SCHOOL
INDUSTRIAL TRAINING INSTITUTE
HEALTH TRAINING AND RESEARCH INSTITUTE

Map 6.31

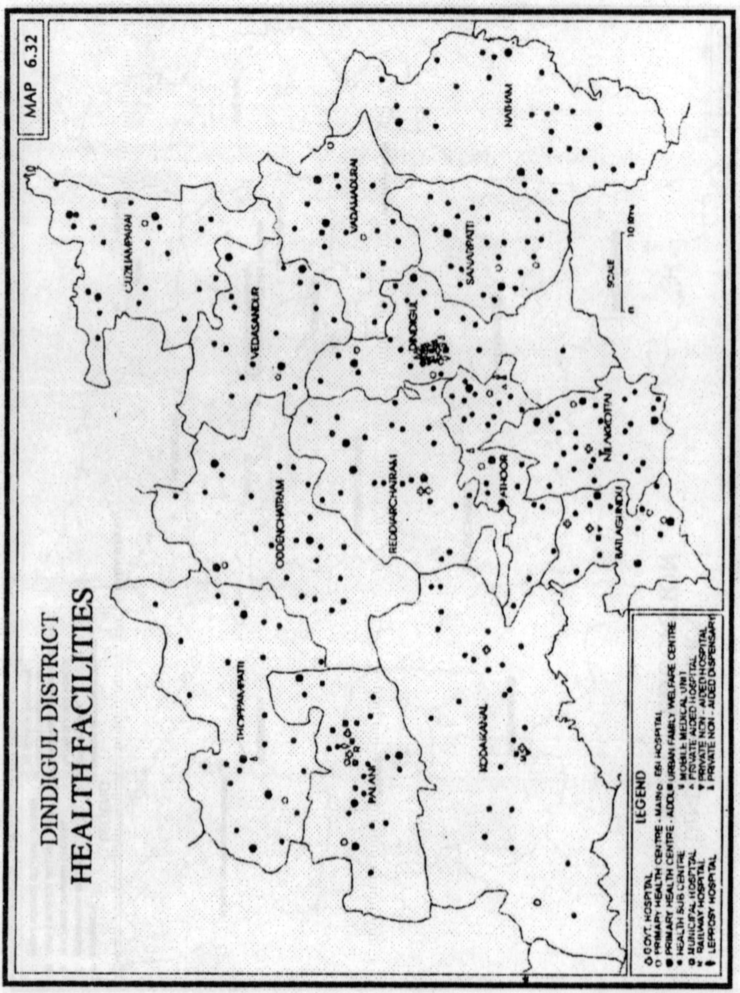
MAP 6.32
DINDIGUL DISTRICT
HEALTH FACILITIES
LEGEND
SCALE
NATHAM
VEDASANDUR
DINDIGUL
KODAIKANAL
PALANI
NILAKKOTTAI

Map 6.32

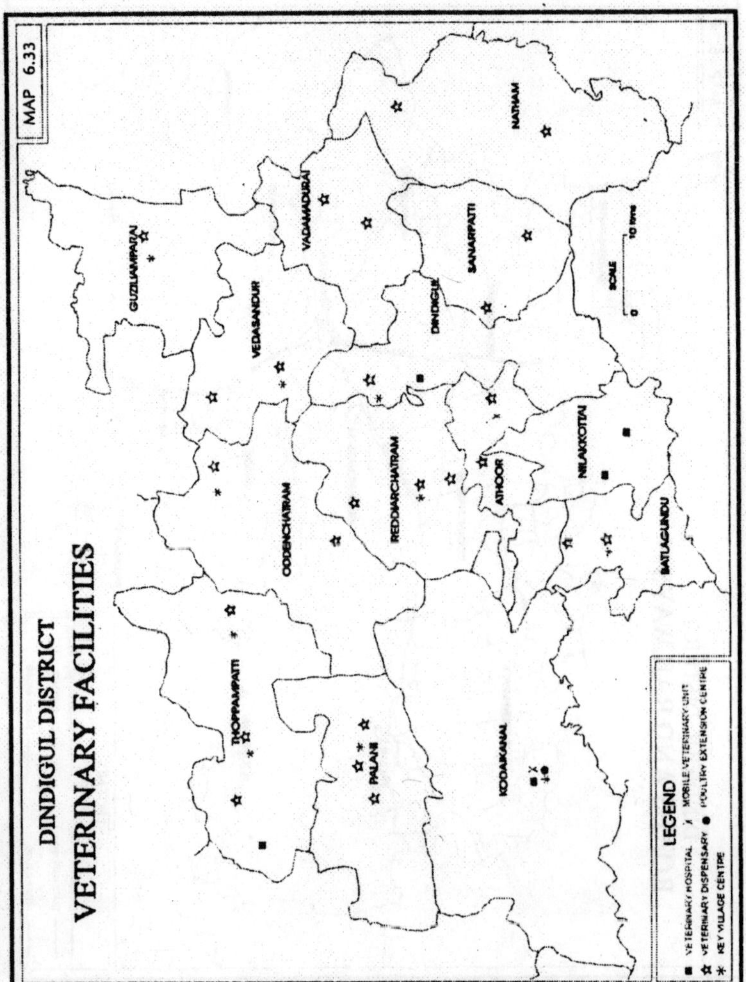
MAP 6.33
DINDIGUL DISTRICT
VETERINARY FACILITIES
GUZILIAMPARAI
VEDASANDUR
VADAMADURAI
NATHAM
SANARPATTI
DINDIGUL
OODENCHATRAM
REDDIARCHATRAM
ATHOOR
NILAKKOTTAI
BATLAGUNDU
THOPPAMPATTI
PALANI
KODAIKANAL
SCALE
0
10 kms
LEGEND
VETERINARY HOSPITAL
VETERINARY DISPENSARY
KEY VILLAGE CENTRE
MOBILE VETERINARY UNIT
POULTRY EXTENSION CENTRE

Map 6.33

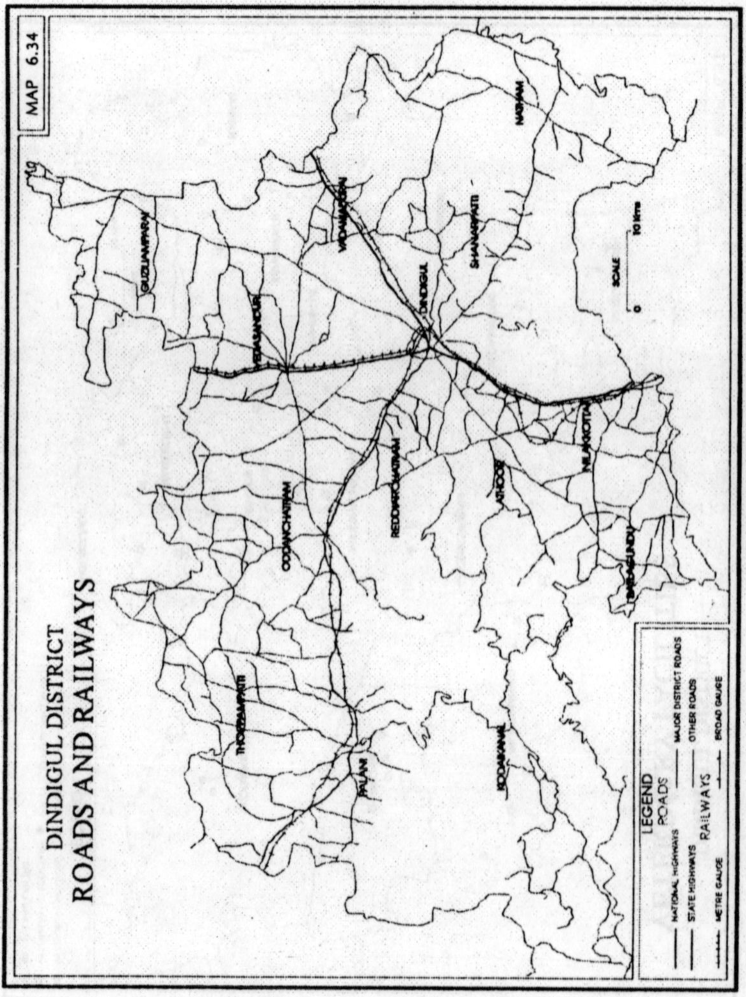
MAP 6.34
DINDIGUL DISTRICT
ROADS AND RAILWAYS
LEGEND
ROADS
NATIONAL HIGHWAYS
STATE HIGHWAYS
MAJOR DISTRICT ROADS
OTHER ROADS
RAILWAYS
METRE GAUGE
BROAD GAUGE
SCALE
0
10 Kms
GUZILIAMPARAI
VEDASANDUR
VADAMADURAI
DINDIGUL
NATHAM
SHANARPATTI
ODDANCHATRAM
REDDIARCHATRAM
ATHOOR
NILAKKOTTAI
BATLAGUNDU
THOPPAMPATTI
PALANI
KODAIKANAL

6.3.2 Explanatory Notes

Map 6.1: This Map shows the location of Dindigul District in India. It is bound by Periyar district in the North, Tiruchirapally district in the East, Madurai district in the south and Coimbatore district and Kerala state in the West. There are fourteen blocks grouped under six *taluks* *viz.,* Palani, Dindigul, Vedasandur, Natham, Kodaikanal, and Nilakottai. The headquarter of the district, Dindigul town is located in Dindigul block.

SECTION I: This has seven Maps, portraying the natural resources like relief, soils, rainfall, temperature etc.

Map 6.2: This Map depicts the general relief of the area. Contours have been used to represent relief. The altitude of the district, with reference to mean sea level ranges from 200 to 2500 mts. Kodaikanal and Sirumalai hills have an altitude of 2500 and 1200 mts. respectively. Both the hills are highly dissected and they rise suddenly from the adjoining level lands. The slopes are very steep as evident from the closely packed contour lines. The northern part of the district and the axis running from Gujiliamparai to Batlagundu are relatively level lands. Kodaganar river has well pronounced valley only in the upper reaches. The hilly region of the district *i.e.* Kodaikanal and Sirumalai supports plantation crops, while other regions cultivate garden and rainfed crops.

Map 6.3: This Map has been designed on the basis of the preceding map – relief. Went Worth's method of slope analysis has been followed to delimit slope regions. Isopleth technique has been adopted to draw this map. Sudden variation in the slope indicates steepness and gentle variation show the nearly level land. In this district, slope suddenly increases from the northern and central plains towards south west and south east *i.e.,* Kodaikanal and Sirumalai hills.

Map 6.4: This Map shows the type of soils in the district. There are 13 types of soil present in the area. The important type of soil is Irugur soil, which is found in the level lands of northern and central parts of the district. Hilly areas predominantly have hilly, Somayyanur and Vylogam soils. Foot hill region has Palaviduthi soil. Other major soil types are Vylogam, Pilamedu, Palathurai and Dindigul.

Map 6.5: Average annual temperature has been portrayed in this plate. Isotherms have been used to represent the data. The lowest and highest temperature coincide with the altitude of the region *i.e.,* Kodaikanal hills record the lowest temperature (20°C) and plains experience the highest temperature (35°C). The major part of the district lies in the temperature region of 30° to 35°C.

Map 6.6: The distribution of annual average rainfall has been presented in this Map. Details of 50 year average rainfall are portrayed in this plate. Isohytes have been used to show the data. The areas of high rainfall coincide with the hilly areas *viz.* Kodaikanal and Sirumalai. The amount of rainfall received in Kodaikanal region varies from 1000 to 1500 mm, while in Sirumalai region it is between 1000 and 1200 mm. The adjoining level lands recorded 600 to 1200 mm of rainfall.

Map 6.7: Map on seasonal rainfall depicts season-wise *viz.*, south east monsoon, North East monsoon, winter and summer rainfall by isohytes. This district receives maximum rainfall during north east monsoon season and it ranges from 400 to 600 mm and South West monsoon supplies 200 to 500 mm. Other seasons *viz.*, summer and winter receives very meagre rainfall ranging from 40 to 300 mm.

Map 6.8: This Map shows the pattern of streams, tanks using qualitative line and point symbols. The additional information *i.e.*, watershed details have been represented by using another qualitative line symbol. The district is dominated by Kodaganar river basin, which has about 40 per cent of the area. Most of the streams originate either in Kodaikanal or Sirumalai hills. Another important river basin is Shanmuganadi. The dentritic pattern (branching like a tree) of streams dominate the river system of this district. The number of tanks is more in these river basins than rest of the water sheds of the district.

SECTION II: This section has three Maps depicting the ground water and irrigation particulars of the district.

Map 6.9: The depth of ground water table is the theme of this Map. Isopleths have been used to represent the depth of water table ranging from 10 to 20 mts. A part of the district, bordering Kodaikanal hills, has a depth of 10 mts. The foot hill region of Sirumalai hills has a depth of 15 mts. Average depth of ground water level in northern and central level land is 15 mts. There are small pockets of less than 10 mts depth in the central and north eastern parts of the district. A few dry pockets are located in Vedasandur and Dindigul *taluks*.

Map 6.10: Ground water quality is portrayed in this Map by employing isopleth technique. The quality of underground water has been classified into four categories *viz.*, good, moderate, fair and inferior quality. It is clear from the map that, northern and central level lands have good to moderate quality of water. The good quality water is available in selected pockets of Kodaganar and Shanmuganadi watersheds. The border areas of the district have only inferior quality of water. Area under inferior quality water is relatively small.

Map 6.11: This Map comprises four maps in one plate. First map depicts the total irrigated area and other maps source-wise (well, tank and other sources) irrigated area. First map drawn with proportionate pie-diagram and other maps by choropleth technique. Oddenchatram block has a large area under irrigation, while Vedasandur has the lowest irrigated area. Predominant source of irrigation in all the blocks is well except in Kodaikanal. It is followed by tank irrigated area particularly in Palani block. In Athoor block, other sources irrigate more area. The second map shows percentage of well irrigated area. In the North East and eastern part of the district, wells are important source of irrigation. Tank irrigated area is more in Shanmuganadi and Kodaganar water sheds.

SECTION III: This section on population has four Maps showing density of population, agricultural workers, literates and vital statistics.

Map 6.12: This Map has been designed to present the density of population by choropleth technique. Athoor, Batlagundu and Palani blocks have the highest density of above 350 persons per sq. km. The lowest density of less than 210 persons per sq. km. is in Gujiliamparai, Kodaikanal and Thoppampatti blocks.

Map 6.13: This Map depicts two sets of data *i.e.,* Occupational structure and percentage of non-workers to total population. The first set of data is represented by pie-diagram and the second one by choropleth technique. Athoor, Palani and Oddenchathram blocks have a higher percentage of non-workers. Non-workers are concentrated in the blocks located in Shanmuganadi, Kodaganar and Vaigai river basins and Natham block. In all blocks, agricultural labourers and cultivators are the largest group followed by other workers, marginal workers and workers in household industries.

Map 6.14: In this Map, data on percentage of literates to the total population are presented by choropleth technique. More than 39 percentage of literates are present in Palani, Oddenchathram and Nilakottai blocks. The blocks located in level lands have more than 30 percentage of literates except Gujiliamparai block. Low percentage of literates *i.e.,* less than 30 per cent are in Kodaikanal, Natham and Gujiliamparai blocks. First two blocks are located in hilly region and the third one in a dry and backward region.

Map 6.15: It has been designed to show birth, death and infant mortality rates by using multiple bar diagrams. Birth rate is higher (more than 30/1000 population) in Palani and Natham blocks than in others. The lowest is reported in Oddenchathram block. Death rate is relatively low in all blocks except Palani block, where it is about 15/1000 population. The highest infant mortality rate (more than 80/1000) is recorded in Palani, Batlagundu and Nilakottai blocks. Palani block dominates with high birth, death and infant mortality rates.

SECTION IV: This section on land use and agriculture has 12 Maps. They show land use, area under crops, landholdings and infrastructural facilities for agriculture.

Map 6.16: This Map portrays the total geographical area of the block and area under different land use types by using proportionate pie-diagram. The biggest block in geographical area is Thoppampatti and the smallest is Batlagundu. Net sown area is the largest component of the total area in Thoppampatti, Palani, Oddenchathram, Reddiarchathram, Athoor, Batlagundu, Nilakottai, Natham and Sanarpatti blocks. Barren and uncultivable lands and land put to non-agricultural uses are more in Kodaikanal *taluk.* 'Other fallows' is the largest component of the area in Gujiliamparai and Vadamadurai blocks.

Maps 6.17 and 6.18: These two (four-maps-in-one) Maps show percentage of different land use types to total geographical area by choropleth technique. Barren and uncultivable land is high in Kodaikanal, Oddenchathram and Reddiarchathram blocks and to a smaller extent in Dindigul, Athoor, Nilakottai and Vedasandur blocks. Cultivable wastelands are found in Oddenchathram, Vedasandur and Dindigul blocks and to a lesser extent in Gujiliamparai, Athoor and Nilakottai blocks Vedasandur, Dindigul, Sanarpatti blocks have above 25 per cent of area under current fallows. Dindigul block has no area under other fallows. In Thoppampatti, Gujiliamparai and Vadamadurai blocks, other fallow lands constitute more than 23 per cent. Athoor, Batlagundu blocks have no area under permanent pasture and other grazing lands. Blocks, located in northern part of the district have more area under permanent pasture and other grazing lands. Land put to non-agricultural uses is relatively large in Palani, Kodaikanal and Sanarpatti blocks. Land under miscellaneous tree crops is relatively high in Oddenchathram and Natham blocks. No area is reported under this type of land use in Reddiarchathram, Athoor and Batlagundu blocks. In Reddiarchathram, Natham and Athoor blocks more than 59 per cent of area remain as net sown area. The lowest area (less than 36 per cent) is found in Vedasandur, Vadamadurai and Gujiliamparai blocks.

Map 6.19: This Map has been designed to show the spatial distribution of forest area. Two sets of data have been presented in this Map *viz.*, reserved forest and reserved land by using chorochromatic technique in colour. The area under forest coincides with hilly region of the district.

In Kodaikanal, majority of the area is under forest. In Sirumalai and Natham areas, forests are distributed in small patches. The reserve lands are in two big patches located in Kodaikanal and Sirumalai hills.

Map 6.20: This Map has four maps showing area under four important crops *viz.*, Bazra, Sorghum, Paddy and Groundnut. Choropleth technique has been used for drawing these maps. Area under Bazra is the highest in Thoppampatti, Vadamadurai, and Natham blocks and no area is reported in Vedasandur and Kodaikanal blocks. Bazra is grown in the central part of the district covering Oddenchathram, Dindigul, Reddiarchathram, Athoor, Nilakottai and Batlagundu blocks. Paddy is an important crop (more than 16.5 per cent of area) in Palani, Nilakottai and Batlagundu blocks. Vedasandur, Gujiliamparai and Oddenchathram blocks have about 14 per cent of area under Groundnut.

Map 6.21: The availability of land (net sown area, current fallows and other fallows) per head of population is presented in the Map. In Dindigul and Palani *taluks*, the availability of land is high, compared to others. The lowest is recorded in Natham and Kodaikanal blocks. Intensity of cropping *i.e.,* percentage of cropped area to net sown area is depicted in another map of this plate. Both these maps are drawn by choropleth technique. Cropping intensity is relatively high in Gujiliamparai, Vedasandur and Natham blocks and medium in Palani, Kodaikanal, Athoor and Dindigul blocks.

Map 6.22: This Map comprises two-maps-in-one Map. The number and area of land holdings are presented in this Map by proportionate pie diagram. Map on number of landholdings shows Sanarpatty, Reddiarchathram and Thoppampatti blocks having the highest number of landholdings. In all these 14 blocks, less than 1 hectare landholdings are more in number closely followed by 1 to 3 hectares. Other categories are relatively negligible. In the map on area of landholdings, Thoppampatti, Oddenchathram and Nilakottai blocks have the largest area of landholdings. Area of landholdings in the size class of less than 1 hectare and 1 to 3 hectares are more in all the blocks than in others. The higher size classes claim about half of the total area of landholdings in all these blocks.

Map 6.23: Availability of agricultural machinery and implements are shown in this Map by using multiple bar diagram. The tractors are more in number in Oddenchathram and oilghanis are more in Batlagundu blocks than in others. In Natham and Oddenchathram blocks, sugarcane crushers are found in small numbers. In majority of the blocks, tractors and sugarcane crushers are not available except in Oddenchathram. Nilakottai block has the largest concentration of ploughs (30000 nos) and Athoor Block the lowest (3000 nos). In Oddenchathram block, about 6000 water pumps are functioning. The least number of machineries and implements are reported in Kodaikanal block.

Map 6.24: In this Map, data on five types of infrastructure facilities are presented in qualitative point symbol. There are 12 main depots and 7 sub-depots of agricultural extension centre in the district. Palani, Oddenchathram, Dindigul and Natham towns have one commercial grading centre each. One agmark lab is located in Dindigul town and seed processing units are in Palani and Dindigul towns. A parasite breeding centre is functioning in Nilakottai town. All the above facilities except parasite breeding centre are available in Dindigul town.

Map 6.25: Location of markets and cattle shandies are shown in this Map. These markets have been represented by qualitative point symbol. In this district, fairs are held in eight places. Regulated markets are functioning at Palani, Oddenchathram, Dindigul, Sanarpatti, Natham and Batlagundu towns. Permanent markets are located at Palani, Vedasandur, Natham, Batlagundu and Dindigul towns. Unregulated markets are more in Sanarpatti block than other parts of the district. Cattle shandies are periodically organised in Oddenchathram, Vadamadurai, Dindigul, Sempatti, Batlagundu and Natham towns. Batlagundu town has all these five types of markets.

SECTION V: Three Maps of this section show industries by multiple bar diagram.

Map 6.26: This Map shows nine groups of industries. Agro-based industries function in all blocks, particularly in Palani and Dindigul blocks. Though forest based industries are located in all blocks (except in Vadamadurai block), the number of units are relatively high in Dindigul block. Animal based units are present only in Thoppampatti, Oddenchathram, Reddiarchathram, Athoor, Gujiliamparai, Vedasandur and Dindigul blocks. Among these blocks, Dindigul block has the largest concentration. Palani has seven textile based industries, 13 chemical based industries and 13 building material industries. Engineering and allied units are concentrated largely in Palani, Dindigul and Athoor blocks. Electrical and miscellaneous industrial units are located in smaller numbers in all the blocks, while Palani block has six units.

Map 6.27: Data on agro-based industries have been depicted in this Map. There are ten groups of industries functioning in this district. Dindigul block dominates with maximum number of coir rope making, mat weaving, vermicelli making units than other blocks. Rice mills are located only in Palani, Gujiliamparai, Vedasandur, Reddiarchathram and Athoor blocks. All blocks have flour mills. Oil mills are concentrated in Palani and Dindigul blocks than others. Palani, Thoppampatti, Oddenchathram, Kodaikanal, Batlagundu, Dindigul and Natham blocks have coffee roasting units. The predominant pickle making centre is Dindigul and important tobacco processing blocks are Palani and Natham. Confectionery units are located in Dindigul, Natham, Sanarpatty, Nilakottai, Palani and Thoppampatti blocks. Dindigul block has more number of agro-based industries than others.

Map 6.28: The district has seven types of artisan-based industries. Silver jewellery making units are present in all blocks except Reddiarchathram, Vedasandur and Kodaikanal. All blocks have units making bamboo products, particularly they are large in number in Gujiliamparai, Batlagundu, Palani and Thoppampatti blocks. Stone sculpture is the predominant activity in Thoppampatti block and art metal ware making units in Palani block. Jacquard looms are operating in Palani, Kodaikanal, Vedasandur, Batlagundu and Dindigul blocks. Units making decorative products are located in Kodaikanal, Athoor and Palani blocks. In Vadamadurai block, gem cutting is an important activity. All these artisan based industries are functioning in Dindigul block.

SECTION VI: This section on infrastructure facilities has five Maps on banks, cooperative societies, schools and colleges, health, veterinary and roads and railways.

Map 6.29: This Map shows the location of financial institutions in Dindigul District. Four groups of banks have been presented by quantitative point symbols. The branches of State Bank of India are located only in western and south western parts of the district. Cooperative banks are concentrated in central part and private banks are more in northern part of the district. Nationalised banks are distributed all over the district. The important banking centre is Dindigul town, having maximum number of all the four types of financial institutions.

Map 6.30: This Map portrays the spatial distribution of cooperative societies by qualitative point symbols. Farmers service cooperative society is functioning only at Ayyampalayam, Palani, Dindigul and Kodaikanal towns. Weavers cooperative societies are concentrated at Chinnalapatti and Dindigul towns. Fishermen cooperative society is located at Palani, Dindigul and Nilakottai towns. Primary Agricultural Credit Banks are uniformly distributed in the district.

Map 6.31: This Map has been designed to depict the distribution of schools, colleges and other educational institutions. High and higher secondary schools are distributed through out the district. The district has been divided into two educational districts. Palani educational district consists of Kodaikanal, Palani and Vedasandur *taluks,* while Dindigul district comprises Nilakottai, Dindigul and Natham *taluks*. Art and Science colleges for men and women are located in Palani and Dindigul towns. Dindigul town has two engineering colleges and two industrial training institutions. A Rural University and a teacher training college are functioning at Gandhigram. Health training and research centres are located at Ambilikkai and Ambathurai.

Map 6.32: Health facilities have been represented in this Map in qualitative point symbols. Primary Health Centres (PHC) are located in all blocks. PHC and PHC sub-centres are evenly distributed except in hilly regions. South western part of the district has majority of the government hospitals. Municipal, leprosy, railway and ESI hospitals are functioning at Palani and Dindigul towns. Private non-aided hospitals are available at Kodaikanal, Batlagundu, Nilakottai and Dindigul towns. Mobile medical facilities are found in Kodaikanal town. Dindigul, Palani and Batlagundu towns have one urban family centre each. All the above health facilities (except mobile facility) are available in Dindigul town.

Map 6.33: Veterinary facilities are presented in this Map by qualitative point symbols. Veterinary hospitals are located at Dindigul and Kodaikanal towns. Veterinary dispensaries are more in plains than in hilly areas. Key village centres are functioning with veterinary dispensaries. Key village centres and sub-centres are distributed only in the northern part of the district. All these veterinary facilities (except key village sub-centre) are located on important roads.

Map 6.34: This Map shows the network of roads and railways by qualitative line symbols. Northern and central part of the district has fairly well developed road network than in other parts. Metre gauge line connects Dindigul town with Palani, Tiruchirapalli and Madurai city.

The recently inaugurated broad gauge line links Dindigul town with Karur town and Madurai city.

The model district atlas, emerging from the present study, thus contain 339 themes drawn from various sources like published atlases, guidelines issued by Planning Commission for district level planning and perception of maps user and designers. The atlas embodies a number of mapping techniques like choropleth, isopleth, chorochromatic etc. The scale of map is 1:250000 and the level of data is block. The data in the map will have to be revised once in five years. Though the themes have been finalised, the list of themes cannot be exhaustive when applied to any particular district, but it can serve as a reliable check list for those preparing maps for district level planning.

REFERENCES

1. T.N. Somaiah., 'Cartography in the Development of District Level Planning', *Indian Cartographer,* Vol. V, 1985, p. 66.
2. V.C. Joshi., 'Village Maps for Rural Development', *Indian Cartographer,* Vol. VIII, 1988, p. 64.
3. P.V. Ganesh., 'District Maps for Developmental Plans upto Block Level', *Indian Cartographer,* Vol. VIII, 1988, p. 110.

7

SUMMARY AND CONCLUSION

7.1 STUDY DESIGN

The present study on 'An Exercise in Perceptional Cartography – Evaluation of Resource Atlases and their Application to District Planning', aims at evolving a list of map themes for District Level Planning.

7.1.1 Objectives

The researcher has sought to accomplish the above task with the following objectives :

1. to identify the sources of data for district planning and problems therein,
2. to study the existing information system at the district level,
3. to evaluate selected resource atlases and guidelines on district planning issued by the Planning Commission, Government of India,
4. to portray the perceptions of map users and views of map designer,
5. to evolve a list of map themes for district level planning on the basis of evaluation of resoure atlases and the percecption of map users and designers, and
6. to prepare a set of model resource maps on selected themes for Dindigul District, Tamil Nadu.

7.1.2 Research Design

The problem associated with sources of data and information system for district planning was studied with the use of pre-tested interview schedule. One national, two state and one district level resource atlases were evaluated for arriving at a list of map themes for district planning. The list was circulated among map users along with a pre-tested interview schedule for getting their perceptions. The same list and another list of maps required by the map users were despatched with a pre-tested mailed questionnaire to the map designers for eliciting their views on the lists and also on various aspects of district resource atlas. Based on the perception of map users and views of map designers, a final list of map themes was prepared with suitable scale, mapping techniques, level of data and periodicity of revision. Further, as an outcome of the study, a set of plates have also been prepared by the researcher as model resource atlas with explanatory notes for Dindigul District, Tamil Nadu.

7.2 Study on Sources of Data and Information System

A review of the various sources of data available for district planning and problems encountered by officials in collection, compilation and use of data has encompassed the sources of data.

This study reveals that the data which are now available for district level planning are gathered by the map users who are working at block, *taluk* or village level. They are saddled with many responsibilities — primarily the implementation of various welfare programmes, besides, their routine administrative work. Further, they have not undergone any training or orientation in scientific method of data collection.

Different departments at the district level collect data at various levels — *Panchayat,* Revenue Village, Block, *Taluk* and District. Some of the departments are acting only as a mere data collecting agency, without knowing either the use of the data or the method of processing the data. Consequently, the reliability and compatibility of official data have suffered considerably.

Each department collects data relating to its work and therefore the data on any sector are not available at one place. Whereas the District Statistical Office collects and computes the data on various sectors of economy based on the details gathered from different offices at various level/time. Further collection, processing and publication of data takes more time and thus when the data made available in hard copies it almost outdated for any planning exercise.

After studying the various problems associated with data sources, it was found imperative to standardise the processes and procedures of data collection, processing and supplying them to the district planners. In this regard, organisations like NIC, DISNIC etc., have been studied with a view to analyse their present function as well as their potential uses as purveyor of data for District level planning in future.

Reluctance on the part of various departments of the State Government in using these facilities and also the multiplicity of level of data, which are now available at district level, are found to have hampered the functioning of these new organisational facilities in information system at district level.

7.3 EVALUATION OF RESOURCE ATLASES

Having studied the data sources and information system available at district level, evaluation of the following resource atlases have been undertaken:

- National Atlas of India,
- Basic Resource Atlas of Tamil Nadu,
- Resource Atlas of Kerala, and
- Planning Atlas of Medak District.

Besides these, the guidelines on district planning issued by the Planning Commission, Government of India in 1969, 1984 and 1990 were examined. The above exercises enabled the researcher to identify a set of 258 map themes for district level planning.

7.4 PERCEPTION OF MAP USERS AND DESIGNERS

The identified 258 map themes which includes 49 maps were circulated among the map users/map designers and their perceptions and views ascertained. The users selected 100 of the 258 map themes and in addition, suggested 75 map themes for inclusion as required for district level planning.

The map users, preferred *colour maps* to black and white maps. Most of them found it easy to grasp the information presented through statistical diagram. Other techniques in the order of their preference were Chorochromatic, Choropleth and Isopleth.

Maps in the *scale of 1:250000* in the paper size of model map were found to be handy for most of them.

Majority of them preferred *block level data* in the maps with facilities for map drawing. They also expressed their need for trained personnel in this regard.

Almost all the departments desired to prepare and use maps and they specified their requirements for necessary facilities including training for personnel in map drawing.

Both the original list of 258 themes as well as the 75 themes required by the users were circulated among the map designers for their views. The designers accepted the original list of 258 map themes in full and also 50 out of 75 themes required by the map users. Further, they themselves suggested an addition of 31 maps. Thus, the list of maps to be prepared at the district level has increased to 339.

Further, the researcher sought to resolve the issue of the *periodicity* of *revision* from both the map users and designers. It was reported that since plans are drawn for a period of five years, revision of maps can be undertaken once in five years and therefore, *quinquennium* can be the periodicity of updating the data required for district level resource maps.

Having finalised the themes of map, level of data, scale, and periodicity of updating, the study proceeded to determine the *techniques of mapping* for 339 map themes of the district resource atlas. In the light of the review of national, state and district level atlases, it was decided by the researcher in tune with the views of the map user and map designer, techniques like Isopleth, Chorochromatic and Choropleth have been chosen for presenting data in the resource maps for district planning.

7.5 MODEL RESOURCE ATLAS

The Researcher re-drew the 49 maps circulated among the map users into 34 Maps. Further, based on the analysis of the perception of the users in all choropleth plates number of classes fixed at four. In addition a few maps were prepared with help of computer. Besides as suggested by the map users relevant explanatory note has been appended to the plates.

For want of time and financial support a selected Maps have been drawn to portray details relating to natural resources, groundwater, irrigation, land use, cropping pattern, industries and infrastructural facilities.

7.6 CONCLUSION

In Indian context, the planning machinery has its apex body at the national level (New Delhi) at the state level (respective state capitals) and at micro-level (respective district headquarters). The national plan is prepared based on the consolidation of information gathered from micro-level (district) but the allocated funds from the Government flow from top to bottom. Under these circumstances, after a careful evaluation of resource atlases at various levels and ascertaining the perception of map users and designers about district level resource atlas, the following suggestions are found necessary :

1. In order to streamline and help planners at the national and state level, a nodal agency like NICNET may supply the data gathered by them from various Government departments and research institutions.
2. It is essential to have standardised Resource Atlas at district level for successful plan formulation and implementation, taking into consideration the micro-level problems and requirements.
3. For comparative study and rational use of resources, it is necessary to standardise the theme, scale, level of data, techniques of mapping and periodicity of updating the resource atlas for district level planning.

Besides the above, in order to highlight certain local problems and requirements based upon local resources, a few additional maps may be added on such of the themes considered essential for the respective micro-regions. Thus, the present study is able to demonstrate the use of cartographic products in decentralised district planning.

APPENDICES

APPENDIX — I

COLLECTION OF DATA AT DISTRICT LEVEL

SOURCES AND PROBLEMS

Interview Schedule

Schedule No. : Date:

1. General Particulars

1.1 Name of the department :

1.2 Name and designation of the Presiding Officer :

1.3 Office address :

2. Sources of Data

2.1 Primary Source
Does your department collect the data directly ? : Yes/No
(*i.e.* primary source)

2.2 If yes, who collects these data sets? :

2.3 Have they undergone training/ orientation in data collection : Yes/No

2.4 If yes, where :

2.5 For how many days/months? :

2.6 What is the level of your data?
1. *Panchayat*
2. Revenue Village

3. Block
4. *Taluk*
5. District
6. Any other (specify)

2.7 How far these data are reliable?
1. To some extent
2. To a moderate extent
3. To a large extent
4. Fully

3. Secondary Sources

3.1 Do you collect data from secondary source? Yes/No

3.2 If yes, what are your sources of secondary data?

3.3 How often you obtain data from these sources?

3.4 Do you find the data adequate? Adequate/Inadequate

3.5 Are you able to collect the latest data? Yes/No

4. Processing of Data

4.1 Do you have a computer for processing the data? Yes/No

4.2 Do you use computer facilities of National Informatics Centre? Yes/No

4.3 If yes, is the computer used to
1. Process the raw data
2. Prepare the report
3. Draw diagrams and graphs
4. Draw maps
5. Any other (specify)

4.4 If no, what are the reasons?

APPENDIX — II

LIST OF MAPS DRAWN BY THE RESEARCHER

1. Location of Dindigul District
2. Relief
3. Slope
4. Soils
5. Annual rainfall
6. Seasonal rainfall
7. Average annual temperature
8. Drainage
9. Depth of ground water table
10. Ground water quality
11. Density of population
12. Working population (as percentage of total population)
13. Agricultural worker (as percentage of total population)
14. Literates (as percentage of total population)
15. Birth, death and infant mortality rates
16. Irrigated area (as percentage of total cropped area)
17. Well irrigated area
18. Tank irrigated area
19. Canal irrigated area
20. Area irrigated by other sources
21. Intensity of cropping
22. Man-land ratio
23. Area of land holdings
24. Number of land holdings
25. Land use
26. Barren and uncultivable land (as percentage of total geographical area)
27. Land put to non-agricultural uses (as percentage of total geographical area)
28. Area under cultivable waste (as percentage of total geographical area)

29. Permanent pastures and other grazing land (as percentage of total geographical area)
30. Land under miscellaneous tree crops (as percentage of total geographical area)
31. Area under current fallows (as percentage of total geographical area)
32. Other fallow lands (as percentage of total geographical area)
33. Net sown area (as percentage of total geographical area)
34. Area under Sorghuns
35. Area under Groundnut
36. Area under Paddy
37. Forest
38. Markets
39. Agricultural machinery and implements
40. Infrastructural facilities for agriculture
41. Industries
42. Agro-based industries
43. Artisan-based industries
44. Cooperative societies
45. Financial institutions
46. Centres of education
47. Veterinary facilities
48. Health facilities
49. Roads..and railways

APPENDIX — III

USE OF MAPS IN DISTRICT LEVEL PLANNING

PERCEPTIONS OF MAP USER

Interview Schedule

Schedule No. : Date:

PART I

General Particulars:

1.1 Name of the Department :

1.2 Year of establishment :

1.3 Name and designation of the Presiding Officer :

1.4 Office address :

1.5 Jurisdiction :
1. Block
2. Division
3. Region
4. District
5. Others (specify)

1.6 Please specify the nature of major activities undertaken by your organisation:
1. Planning
2. Implementation & Monitoring
3. Other activities (specify)

Use of Maps

1.7 Do you use map in day-to-day activities of your department? Yes/No

1.8 If yes, please specify in which of the following activities you use maps?

1. Planning
2. Implementation & Monitoring
3. Other activities (specify)

1.9 If no,

1. List the reasons for not using the maps?
2. What do you require to increase the use of maps?

1.10 List the theme of maps used in the following activities:

Sl. No.	Nature of Activity	List of Maps
1.	Planning	
2.	Implementation and Monitoring	
3.	Other activities	

1.11 From where do you get the above Maps?

1. Prepared by the department
2. Borrowed from other departments
3. Purchased from other departments
4. Others (specify)

1.12 Were these maps drawn to scale?
Yes/No

1.13 If yes, what is the scale of the map?

1.14 What are your suggestions to increase the use of maps?

PART II

Perception of Map Users about Maps/Atlas

(Respondents will be shown a set of maps on various themes, related to Dindigul District, prepared by the researcher in order to know about the perception)

2.1 Do you find it difficult to understand the data presented in these maps? Yes/No

2.1.1 What is your order of preference of the following mapping techniques in understanding the map themes?
1. Isopleth - Pattern
2. Chorochromatic - Pattern
3. Choropleth - Pattern
4. Diagram - Pattern

2.1.2 What are your suggestions for modifying the method of presentation of data in the map?

2.1.3 What is the most suitable scale of the map?

2.1.4 Do you prefer computer drawn maps in the place of hand drawn maps. Yes/No

2.1.5 Do you require a map drawn based on :
1. Revenue Village Level Data,
2. Block Level Data.

2.1.6 Please select maps for your departmental use from the following maps/list of themes (for List of themes, Please see table 4.5).

2.1.7 Please suggest additional theme of maps required by your Department.

2.2 If an atlas is prepared for the purpose of planning for your Department, please specify your preferences.

2.2.1 How often this planning atlas should be revised?

1. Every Year
2. Every Two Years
3. Every Three Years
4. Every Five Years
5. Every Ten Years

2.2.2 What should be the scale of this map?

1. 1:16000
2. 1:50000
3. 1:250000
4. Others (specify)

2.2.3 Please suggest the lowest level for which data is required for drawing maps?

1. *Panchayat*
2. Revenue village
3. Block
4. *Taluk*
5. Any other (specify)

2.2.4 Would you like to have an explanatory note to supplement the map? Yes/No

2.2.5 Do you require the Atlas in

1. Black and white
2. Colour

2.2.6 Would you like to use a computer for the purpose? Yes/No

2.2.7 Do you have any suggestions in this regard?

APPENDIX — IV

USE OF MAPS IN DISTRICT LEVEL PLANNING

VIEWS OF MAP DESIGNERS

Questionnaire

Date :

Schedule No. :

PART I

Sl. No.	Theme of maps	Tick appropriate map for District planning	If the map is incomplete, suggest the required modifications/ additions/ deletions	Remarks
1	2	3	4	5
a)	For list of 258 themes, please see Table 4.5			
b)	For list of 75 themes required by the map users, please see Table 5.6			

PART II

2. Are you satisfied with the list of maps? Yes/No

2.1 If no, suggest additional maps which are required for district level planning

Sl. No.	Theme of maps

2.2 Please tick the most appropriate level of data that could be used for district level planning

1. *Panchayat*
2. Revenue village
3. Block
4. *Taluk*
5. Others (specify)

2.3 What is the suitable scale for district level resource atlas?

Scale :

2.4 Suggest the periodicity of revision of the proposed atlas.

APPENDIX — Va

NATIONAL ATLAS OF INDIA

List of Themes

1. Administration
2. Parliamentary constituency
3. Physical
4. Physiography
5. Physiographic divisions
6. Morthophochronodogy (Geology)
7. Relative relief
8. Slope
9. Depth of soil
10. Type of rocks
11. Type of minerals
12. Occurrence and exploitation of minerals
13. Soil texture (Type of Soil)
14. Soil reaction (pH value)
15. Earthquakes
16. Palaeogeography
17. Annual temperature
18. Wet bulb temperature
19. Seasonal temperature — Jan, Apl, July, Oct
20. Pressure of wind
21. Cloudiness for Jan, Apl, July, Oct
22. Accumulated temperature
23. Isotherm for seed germination
24. Isotherm for human comfort
25. Annual rainfall
26. Seasonal rainfall (Jan-Feb/March-May/June-Sept/Oct-Dec)
27. Monthly rainfall (January to April)
28. Monthly rainfall (May to July)
29. Monthly rainfall (Sept to Dec)

30. Rainfall — monthly rainfall
31. Variation of rainfall
32. Annual rainy days
33. Seasonal rainy days (Jan-Feb/Mar-May/Jun-Sept/Oct-Dec)
34. Percentage of South West Monsoon to total annual rainfall
35. Drainage
36. Monthly runoff
37. Annual runoff
38. Water resources
39. Normal annual rainfall
40. Water table
41. Quality of water, location of dams, canals and command area
42. Indian Ocean
43. Bay of Bengal
44. Wild life & wetlands
45. Forest types
46. Industrial wood (Production)
47. Fuelwood (Production)
48. Minor forest produce (Production)
49. Occupational structure
50. Rural population
51. Density of rural population (1951)
52. Variation of total population
53. Density of rural population
54. Changes in rural population
55. Urban population
56. Density of rural population
57. Transport and tourism
58. Telecommunication
59. Number of subscribers
60. Volume of trunk call
61. Mails carried by air
62. Land use types
63. Growth of industries — 1921, 1931, 1947, 1961
64. Engineering industries
65. Cotton and jute
66. Area under wool and silk
67. Variety of silk
68. Variety of wool
69. Major/Minor sugarcane area

70. Percentage of sugar received
71. Distribution of handicrafts making centres
72. Handloom production
73. Location of handloom centres
74. Production
75. Crop regions
76. Soil fertility (NPK)
77. Rice, millet, wheat
78. Cereals percapita production
79. Millets (6 crops)
80. Production of millet
81. Pulses (3 crops)
82. Production of pulses
83. Percentage of area under sugarcane to total cropped area and production
84. Production of khandasari
85. Production of palmgur and gur
86. Plantation and garden crops
87. Inland movement of tea and coffee
88. Oilseeds
89. Production of oilseeds
90. Production of vanaspati factories
91. Vegetables and root crops
92. Cold storage for potatoes
93. Distribution of livestock
94. Percentage variation of milch cattle and buffaloes
95. Working cattle/10 hec. of arable land
96. Number of sheep, goat and pig/km^2
97. Bovine population/km^2
98. Working animal
99. Milch animal and buffaloes
100. Cattle and buffaloe breeds
101. Stock density
102. Markets
103. Distribution of livestock units
104. Fisheries and poultry
105. Distribution of fisheries and poultry units
106. Percapita annual consumption of fish
107. Annual fish landing
108. Movement of fish

109. Rural markets
110. Spacing of markets
111. Flow of commodities
112. Location of ports
113. Flow of commodities by roads and railways
114. Coastal trade in the ports
115. Forest economy
116. Forest reserves
117. Person/hec of forests
118. Livestock market
119. Fair/market
120. Movement of sheep, cattle and goats
121. Agricultural markets
122. Annual arrival of important food grains
123. Location of markets for sugarcane
124. Lac and tobacco
125. Land under cultivation
126. Presure on land
127. Land ownership
128. Agricultural landholdings
129. Non-agricultural areas
130. Available/stock farming/forest
131. Banking facilities in relation to population
132. Bank deposits
133. Co-operative banks
134. Insurance
135. Business transactions
136. Non-Indian/Indian insurers
137. Land capability
138. Cholera
139. Tuberculosis hospital
140. Leprosy — number of beds
141. Cancer death/1000 population
142. Spacing of villages
143. Crowding of houses
144. Crowding of houses in urban areas
145. Density of houses
146. Density of houses in cities
147. Literacy
148. Literacy in selected cities

149. Languages
150. Percentage of school age-group children to total population
151. Location of universities and strength
152. Location of facilities for technical education
153. Percentage of technical literates
154. Location of urban centres according to population
155. Changes in population
156. Age — sex : changes in population
157. Age — sex structure
158. Health
159. Number of beds in hospitals
160. Number of persons/hospital bed
161. Area of water supply, canal, tank, tubewell etc.
162. Water for domestic and industrial use
163. Domestic water supply per day/head of urban population
164. Rural settlement pattern
165. Landholding : Marginal
166. Small
167. Semi medium
168. Medium
169. Large
170. Agricultural labour
171. Work force
172. Forest — reserved forest, protected forest and others
173. Arable/grass/water features
174. Historical – monuments
175. Indo-Aryan style monuments
176. Ancient Kingdom of Kings-temples
177. Medieval
178. India in 1836, 1857, 1919-1947, 1956
179. Archaeology
180. Agricultural landscape
181. Predominant urban activity
182. Pastoral/fishing landscape
183. Wild landscape
184. Village types
185. Urban land use and density of population

APPENDIX — Vb

BASIC RESOURCE ATLAS OF TAMIL NADU

List of Maps

Map **Title of Maps**

1. Administrative divisions

LAND RESOURCES

2. Relief
3. Geology
4. Soils
5. Slope
6. Land capability

WATER RESOURCES

7. Annual rainfall
8. Seasonal rainfall

   (a) South-West monsoon rainfall (June-September)
   (b) South-West monsoon rainfall as per cent of annual rainfall
   (c) North-East monsoon rainfall (December-February)
   (d) North-East monsoon rainfall as per cent of annual rainfall

9. Variability of rainfall

10. Water surplus and deficiency

    (a) Mean annual actual evapotranspiration
    (b) Mean annual water surplus
    (c) Aridity
    (d) Mean annual water deficiency

11. Cyclonic storms and affected areas

12. Drainage

13. Ground water

    (a) Ground water level
    (b) Ground water quality

BIOTIC RESOURCES

14. Vegetation

15. Forests

16. Forest economy

    (a) Production of timber, sandal wood and pulp wood
    (b) Production of fuel wood
    (c) Minor forest produce
    (d) Forest revenue

17. Animal Husbandry

    (a) Milch animals
    (b) Livestock other than milch animals
    (c) Dairy projects
    (d) Poultry units

18. Wild life and sanctuaries

19. Marine fisheries

    (a) Marine fish production
    (b) Fishing equipments
    (c) Fish meal plants
    (d) Fish processing industries

MINERAL RESOURCES

20. Distribution of minerals
    Inset: Mineral production

HUMAN RESOURCES

21. Distribution of population

22. Birth, death and infant mortality rates

23. Growth of population

    (a) 1961-71 (district-wise)
    (b) 1971-81 (district-wise)
    (c) 1971-81 (*taluk*-wise)
    (d) 1951-81 (district-wise)

24. Density, urbanisation, sex-ratio and migration

    (a) Density of population
    (b) Rural/urban composition
    (c) Sex-ratio
    (d) Migration

25. Net-Migration in selected districts

    (a) Coimbatore
    (b) Madras - Chenglepattu
    (c) Madurai
    (d) Tiruchirapalli

26. Age-structure and literacy

    (a) Age-structure of population
    (b) Illiterate population (over 5 years of age)
    (c) Literacy
    (d) Female literacy

27. Occupational structure

    (a) Work participation
    (b) Agricultural workers
    (c) Industrial workers
    (d) Other workers

IRRIGATION

28. Sources of irrigation

29. Extent of irrigation
Inset: Changes in the extent of irrigated area

30. Irrigated area according to sources

(a) Tank irrigation
(b) Well irrigation
(c) Canal irrigtion
(d) Net-irrigated area

AGRICULTURAL DEVELOPMENT

31. Landuse pattern

32. Cultivated area

(a) Net cultivated area
(b) Intensity of cropping
(c) Man-land ratio
(d) Land holdings of small and marginal farmers

33. Cultivation of Rice and Sorghum

(a) Area under Rice
(b) Area under Rice (as per cent of total cropped area)
(c) Area under Sorghum
(d) Area under Sorghum (as per cent of total cropped area)

34. Cultivation of Bajra and Pulses 1977-78

(a) Area under Bajra
(b) Area under Bajra (as per cent of total cropped area)
(c) Area under Pulses
(d) Area under Pulses (as per cent of total cropped area)

35. Cultivation of Groundnut and Other oilseeds

    (a) Area under Groundnut
    (b) Area under Groundnut (as per cent of total cropped area)
    (c) Area under other oilseeds
    (d) Area under other oilseeds (as per cent of total cropped area)

36. Cultivation of Sugarcane and Cotton

    (a) Area under Sugarcane
    (b) Area under Sugarcane (as per cent of total cropped area)
    (c) Area under Cotton
    (d) Area under Cotton (as per cent of total cropped area)

37. Relative yield and spread indices

    (a) Rice
    (b) Sorghum
    (c) Bajra
    (d) Pulses

38. Relative yield and spread indices

    (a) Groundnut
    (b) Cotton
    (c) Other oilseeds
    (d) Sugarcane

39. Change in area of major foodgrain crops

    (a) Rice
    (b) Sorghum
    (c) Bajra
    (d) Pulses

40. Change in area of major non-food grain crops

    (a) Sugarcane
    (b) Cotton
    (c) Groundnut

(d) Other oilseeds

41. Crop regions

42. Land holdings

(a) Area of land holdings
(b) Number of land holdings

43. Tenent cultivation

(a) Area of land holdings
(b) Number of land holdings

44. Agricultural markets

45. Agricultural credit societies

46. State seed farms

47. Farm mechinery

(a) Bullock and manually operated implements
(b) Tractor and tractor operated implements
(c) Oil engines, electrical pumpsets and sprayers
(d) Sugarcane crushers, oil ghannies and thrushers

INDUSTRIAL DEVELOPMENT

48. Power development

(a) Power generation
(b) Power grid

49. Industrial locations

(a) Food processing and textile industries
(b) Wood, paper and chemical products
(c) Metallic and non-metallic products
(d) Machinery, miscellaneous industries and repairs

50. Industrial production

51. Industrial employment

52. Tourist centres

DEVELOPMENT INFRASTRUCTURE

53. Roads and railways

54. Bus traffic flow

55. Air ways and sea routes

56. Education

 (a) School education districts
 (b) Number of school age-group children per school
 (c) Centres of higher education
 (d) Higher education — strength of students

57. Post and telecommunication services

 (a) Post offices (according to postal divisions)
 (b) Telegraph grid
 (c) Postal and telegraphic services
 (d) Trunk call services

58. Medical services
 Inset: Primaiy health centres

59. Banking services

REGIONAL RESOURCE DEVELOPMENT

60. Levels of regional resource development

 (a) Agricultural development
 (b) Livestock development
 (c) Industrial development
 (d) Regional resource development

APPENDIX — Vc

RESOURCE ATLAS OF KERALA

List of Maps

Plate	Title of Map
1.	Administrative divisions Inset: Location map
NATURAL RESOURCES	
2.	Physiography Inset: Altitudinal zones
3.	Drainage
4.	Slope
5.	Geology
6.	Hydrogeology Inset: Ground water provinces of Kerala
7.	Soil
8.	Climate: Temperature — Mean monthly temperature (1931-'79) Inset: Mean annual temperature (1931-'79)
9.	Climate: Rainfall — Mean monthly and annual rainfall (1901-'79)
10.	Climate: Rainfall — South-West monsoon (1901-'79) Inset: Average number of rainy days (1901-'79)
11.	Climate: Rainfall — North-East monsoon (1901-'79) Inset: Average number of rainy days (1901-'79)

12. Climate: Rainfall — Other than monsoon (1901-'79)
    Inset: Average number of rainy days (1901-'79)

13. Climate: Rainfall (1901-'79)

    (a) Rainfall variability (North-East monsoon)
    (b) Rainfall variability (South-West monsoon)
    (c) Rainfall variability (Dry season)
    (d) Variability of annual rainfall

14. Climate: Wind speed and direction (1979-'80)

15. Vegetation: Forest — Adjacent plantations

HUMAN RESOURCES

16. Population: Rural 1981

17. Population: Urban 1981

18. Population:

    (a) Density 1981
    (b) Decadal variation of density (1971-'81)
    (c) Sex-ratio 1981
    (d) Decadal variation of sex ratio (1971-'81)

19. Population: Scheduled Castes 1981
    Inset: Scheduled Tribes 1981

20. Population:

    (a) Literacy 1981
    (b) Decadal variation of literacy (1971-'81)
    (c) Male literacy 1981
    (d) Female literacy 1981

21. Population: Occupational structure

    (a) Cultivators 1981
    (b) Agricultural labourers 1981

(c) Household industries 1981
(d) Other workers 1981

22. Population:

(a) Occupational structure
(b) Main workers 1981
(c) Male main workers 1981
(d) Female main workers 1981
(e) Marginal workers 1981

INFRASTRUCTURE

23. Irrigation: Command area 1982
Inset: Source-wise irrigation 1982

24. Power: Generation and distribution 1983
Inset: Consumption 1979

25. Communication: Transport network, 1981
Inset: Road accessibility 1981

26. Communication:

(a) Inland water ways, 1983
(b) Railway accessibility 1983
(c) Road density 1981
(d) Road availability 1981

27. Communication: Distribution of post & telegraph offices 1983
Inset: Population per post office 1981

28. Financial Institutions: Distribution of banks 1976
Inset: Population per bank 1981

29. Financial Institutions: Bank advances to agricultural sector 1980

30. Financial Institutions: Distribution of cooperatives 1979

31. Educational Institutions:

    (a) High Schools 1981
    (b) College 1981
    (c) Professional colleges 1982
    (d) Technical institutions 1981

RESOURCE DEVELOPMENT

32. Minerals: Reserves 1976

33. Forest Utilization:

    (a) Fuelwood and matchwood
    (b) Other Indian wood
    (c) Pulpwood
    (d) Plywood

34. Land use: General

35. Land use: Classification

    (a) Forest 1978-79
    (b) Permanent pastures and grazing land 1978-'79
    (c) Land put to non-agricultural uses 1978-'79
    (d) Barren and uncultivable land 1978-'79

36. Land use: Classification

    (a) Land under miscellaneous tree crops 1978-'79
    (b) Cultivable wasteland 1978-'79
    (c) Fallows other than current fallows 1978-'79
    (d) Current fallow 1978-'79

37. Agriculture:

    (a) Net area sown 1973-'78
    (b) Area sown more than once 1973-'78
    (c) Area irrigated 1973-'78
    (d) Cropping intensity 1973-'78

38. Agriculture :

(a) Rice — Area 1973-'78
(b) Rice — Yield 1973-'78
(c) Tapioca — Area 1973-'78
(d) Tapioca — Yield 1973-'78

39. Agriculture :

(a) Coconut — Area 1973-'78
(b) Coconut — Yield 1973-'78
(c) Castor — Area 1973-'78
(d) Castor — Yield 1973-'78

40. Agriculture :

(a) Tea — Area 1973-'78
(b) Tea — Yield 1973-'78
(c) Coffee — Area 1973-'78
(d) Coffee — Yield 1973-'78

41. Agriculture :

(a) Pepper — Area 1973-'78
(b) Pepper — Yield 1973-'78
(c) Betelnut — Area 1973-'78
(d) Betelnut — Yield 1973-'78

42. Agriculture :

(a) Sesame — Area 1973-'78
(b) Sesame — Yield 1973-'78
(c) Pulses — Area 1973-'78
(d) Pulses — Yield 1973-'78

43. Agriculture :

(a) Other Plantain — Area 1973-'78
(b) Other Plantain — Yield 1973-'78
(c) Sugar — Area 1973-'78
(d) Sugar — Yield 1973-'78

44. Agriculture :

(a) Banana — Area 1973-'78
(b) Banana — Yield 1973-'78
(c) Rubber — Area 1973-'78
(d) Rubber — Yield 1973-'78

45.. Agriculture: Cropping pattern 1978

46. Livestock: Percentage distribution of livestock 1977
Inset: Dairy 1979

47. Livestock: Poultry

(a) Distribution of poultry population 1973(1)
(b) Distribution of poultry population 1973(2)
(c) Density of poultry population 1973
(d) Availability of poultry per 1000 population 1973

48. Fisheries:

(a) Marine fisheries 1979-80
(b) Inland fisheries 1979-80
(c) Ice plant, cold storage and canning centres
(d) Fish market centres

49. Industries: Distribution of industrial units 1979

50. Industries: Distribution of industrial workers 1978-79

51. Tourism: Places of tourist importance 1983

REGIONS AND REGIONAL DEVELOPMENT

52. Natural regions

53. Bio-climatic regions

54. Crop combination regions 1978

55. Development regions — Level of Agricultural Development 1979-80

56. Development regions: Levels of Industrial Development 1979-80

57. Development regions : Functional Hierarchy of Urban Centres 1971

58. Development regions: Levels of development 1980

APPENDIX — Vd

PLANNING ATLAS OF MEDAK DISTRICT

List of Maps

LOCATION AND ADMINISTRATIVE DIVISIONS

1. Location of Medak District

2. Medak District Administrative

PHYSICAL SETTING

3. Drainage
   Inset: Relief

4. Climate: Isohytes and monthly rainfall
   Inset: Recharge
   Run off

5. Geology:
   Inset: Lineaments
   Bedrock configuration

6. Depth of water table
   Inset: Water quality: Hardness

7. Soils : Chemical parameter
   Sodium in top soil
   Phosphorous in top soil
   Potassium in top soil
   Chloride in top soil

8. Soils : Classification
   Inset : Thickness of top soil

9. Land use: Physical

10. Land use: General
    Area under forest
    Area under cultivable waste
    Net sown area
    Uncultivable area

11. Index Map: Index to village code

ECONOMIC SETTING

12. Agriculture: Area under Kharif dry crops

13. Area under Rabi crops

14. Area under principal food crops

15. Area under Paddy

16. Area under Jowar

17. Area under Bajra

18. Area under Maize

19. Area under Pulses

20. Area under principal commercial crops

21. Area under Sugarcane

22. Area under Groundnut

23. Area under Castor

ECONOMIC SETTING: IRRIGATION

24. Total area irrigated

25. Irrigated area by sources
    Canals
    Tanks
    Wells
    Other sources

26. Ground water potential

POPULATION

27. Growth rate 1971-81

28. Density of population 1971

29. Density of population 1981

30. Scheduled Castes : Total Scheduled Castes Population 1981
    Inset: Total Scheduled Castes Population

31. Scheduled Tribes

32. Literates, 1981
    Inset: Male literates

33. Workers: Main and marginal workers

    Total main workers
    Male main workers
    Total marginal workers
    Male marginal workers

34. Workers: Agricultural workers

    Total cultivators
    Male cultivators
    Total agricultural labourers
    Male agricultural labourers

35. Workers: Household industry
    Total workers in household industry
    Inset: Male worker in household industry

36. Workers

37. Non-workers

38. Settlements: Distribution of urban and rural settlements

SOCIO ECONOMIC INFRASTRUCTURE

39. Medical

40. Educational

41. Communications (Post & Telegraph)

42. Banking

RURAL DEVELOPMENT PROGRAMMES

43. Roads under Public Works Department, Jilla Parishad and Panchayat Raj Departments
    Inset: Construction of buildings under National Rural Employment Programme

44. Supply of drinking water

45. Rural housing programme

46. Housing for weaker sections

    Scheduled Castes
    Scheduled Tribes
    Backward Classes
    Economically Backward Classes

47. IRDP: Loans advanced to beneficiaries under various schemes

48. Marketing: Location of grain and livestock marketing

    Grains : Market arrivals and values
    Livestock : Market arrivals and values

Grains : Estimated market arrivals and values during 1985-90

49. Power : Rural electrification
Inset : Number of agricultural consumers
Number of industrial consumers

INDUSTRIAL DEVELOPMENT

50. Large and medium scale Industries

Industries : Workers employed and capital invested in large and medium scale industries
Inset : Small scale industries
Distribution of large and medium scale industries (Proposed)

51. Small scale Industries

Agro-based Industries:

Number of units and number of workers
Capital invested and value of annual output

Agricultural Implements and Manufacturing Industries:

Number of units and number of workers
Capital invested and value of annual output

52. Small scale Industries

Wood based Industries:

Number of units and number of workers
Capital invested and value of annual output

Rubber and Leather Industries:

Number of units and number of workers
Capital invested and value of annual output

53. Small scale Industries

Chemical Industries:

Number of units and number of workers
Capital invested and value of annual output

Ceramics and Building Materials Industries:

Number of units and number of workers
Capital invested and value of annual output

54. Small scale Industries

Engineering Industries:

Number of units and number of workers
Capital invested and value of annual output

Electrical and Allied Industries:

Number of units and number of workers
Capital invested and value of annual output

55. Small scale Industries

Printing, Binding and Allied Industries:

Number of units and number of workers
Capital invested and value of annual output

Miscellaneous Industries:

Number of units and number of workers
Capital invested and value of annual output

APPENDIX — VI

LIST OF THEMES BASED ON THE EVALUATION OF ATLASES

Sl. No.	Themes of Map
1.	Location
2.	Administration
3.	Relief
4.	Slope
5.	Drainage
6.	Geology
7.	Soil
8.	Chemical properties of top soil — Sodium
9.	Phosphorus
10.	Potassium
11.	Chloride
12.	Bedrock configuration
13.	Thickness of top soil
14.	Annual rainfall
15.	Monthly rainfall
16.	Seasonal rainfall
17.	Seasonal rainy days
18.	Mean annual actual evapotranspiration
19.	Mean annual water surplus
20.	Aridity
21.	Mean annual water deficiency
22.	Recharge of water
23.	Water runoff
24.	Variability of rainfall
25.	Annual temperature
26.	Monthly temperature
27.	Wind speed and direction
28.	Natural vegetation
29.	Forests

30. Forest utilization: Fuelwood
31. Pulp and other Indian wood
32. Plywood
33. Minor forest produce
34. Forest revenue
35. Depth of water table
36. Ground water quality
37. Ground water potential
38. Hydro-geology
39. Sources of irrigation
40. Extent of irrigation
41. Irrigated Area: Tank
42. Well
43. Canal
44. Other sources
45. Net irrigated area
46. Total irrigated area
47. Changes in the extend of irrigated area
48. Land use
49. Land capability
50. Percentage of area under forest
51. Percentage of barren and uncultivable land
52. Percentage of cultivable waste land
53. Percentage of area under fallows other than current fallows
54. Percentage of area under current fallows
55. Percentage of area under permanent pastures and grazing lands
56. Percentage of area under non-agricultural uses
57. Percentage of area under miscellaneous tree crops
58. Percentage of net sown area
59. Cropping pattern
60. Area under kharif dry crops
61. Area under rabi crops
62. Area and yield of principal food crops
63. Area and yield of principal commercial crops
64. Area under Rice
65. Area under Bajra
66. Area under Jowar
67. Area under Maize

68. Area under Sorghum
69. Area under Pulses
70. Area under Banana
71. Area under Other plantain
72. Area under Topioca
73. Area under Groundnut
74. Area under Sesame
75. Area under Castor
76. Area under Coconut
77. Area under Other oilseeds
78. Area under Sugarcane
79. Area under Cotton
80. Area under Tea
81. Area under Coffee
82. Area under Pepper
83. Area under Betelnuts
84. Area under Rubber
85. Yield of Rice
86. Yield of Bajra
87. Yield of Jowar
88. Yield of Maize
89. Yield of Pulses
90. Yield of Sorghum
91. Yield of Banana
92. Yield of Other plantain
93. Yield of Topioca
94. Yield of Groundnut
95. Yield of Sesame
96. Yield of Castor
97. Yield of Coconut
98. Yield of Other oilseeds
99. Yield of Sugarcane
100. Yield of Cotton
101. Yield of Tea
102. Yield of Coffee
103. Yield of Pepper
104. Yield of Betelnuts
105. Yield of Rubber
106. Changes in the area of Rice
107. Changes in the area of Sorghum

148. Growth of population 1951-81
149. Density of population 1981
150. Density of population 1971
151. Distribution of population
152. Distribution of urban and rural settlements
153. Functional hierarchy of urban centres
154. Total Scheduled Castes population 1981
155. Total Scheduled Castes population 1971
156. Total Scheduled Tribes population
157. Migration
158. Sex-ratio
159. Decadal variation in sex-ratio
160. Literates
161. Male literates
162. Decadal variation in literacy
163. Birth, death and infant mortality rates
164. Rural-urban composition
165. Age-structure of population
166. Occupational structure
167. Percentage of workers
168. Percentage of non-workers
169. Percentage of cultivators
170. Percentage of male cultivators
171. Percentage of agricultural labourers
172. Percentage of male agricultural labourers
173. Percentage of household industries
174. Percentage of male workers in household industries
175. Percentage of main workers
176. Percentage of male main workers
177. Percentage of marginal workers
178. Percentage of male marginal workers
179. Percentage of other workers
180. Construction of buildings under National Rural Employment Programme
181. Supply of drinking water : Sources and Quality of water
182. Rural housing programme
183. Housing for Scheduled Castes
184. Housing for Scheduled Tribes
185. Housing for Backward Classes
186. Housing for economically backward classes

221. Road availability
222. Road density
223. Railway accessibility
224. Inland waterways
225. Airways and sea routes
226. Centres of education
227. School education districts
228. Students strength of centres of education
229. Banks
230. Population served per bank
231. Telegraph grid
232. Post and telegraph offices
233. Population served per post office
234. Hospitals
235. Primary health centres
236. Natural regions
237. Bio-climatic regions
238. Levels of agricultural development
239. Levels of industrial development
240. Levels of livestock development
241. Levels of development

APPENDIX — VIIa

LIST OF MAPS SUGGESTED IN GUIDELINES FOR DISTRICT LEVEL PLANNING - 1969

Sl. No.	Themes
1.	Elevation
2.	General topography
3.	Rainfall
4.	Rivers and streams
5.	Drainage
6.	Depth of underground water strata
7.	Soil structure
8.	Vegetation
9.	Flora and fauna
10.	The composition and distribution of population in terms of rural/urban habitations, locations
11.	The special economic, social and cultural characteristics
12.	Levels of development
13.	Area liable to flooding, water logging/soil erosion/ alkalinity
14.	Cropping pattern
15.	Industrial activities
16.	Communication system
17.	Power distribution system
18.	Location of administrative centres
19.	Rural/Urban, mandi towns
20.	Storage godowns
21.	Primary health centres
22.	Human and Veterinary hospital/dispensaries
23.	Educational institutions
24.	Banking and credit facilities
25.	Irrigation

APPENDIX — VIIb

LIST OF MAPS SUGGESTED IN THE GUIDELINES FOR DISTRICT LEVEL PLANNING - 1984

Sl. No.	Themes
1.	Topographical characteristics including elevations by suitable contours; rainfall distribution — monthly and seasonal; underground water, water strata
2.	Problem areas that are liable to flooding, water logging, soil erosion/alkalinity, recurrect drought
3.	Landuse pattern
4.	Cropping pattern
5.	Sources, location and command of existing irrigation works both private and public
6.	Transport and communication network
7.	Power distribution network
8.	Separate map showing location of - — Educational facilities
9.	— Health facilities
10.	— Veterinary facilities
11.	— Hats and bazars and agro-service centres
12.	— Seed/fertilizer distribution, cooperative, fairs
13.	— Banking and credit facilities
14.	— Flour mills, rice mills, oil expellers and other industrial units
15.	— Minerals
16.	Drinking water facilities showing the sources and quality of water available
17.	Concentration of Scheduled Castes/Tribes and landless agricultural labourers
18.	Distribution of settlements by population size
19.	Separate maps showing projected values of (a) Population (b) Agricultural production (c) Other production levels

APPENDIX — VIIc

LIST OF MAPS SUGGESTED IN THE GUIDELINES FOR DISTRICT LEVEL PLANNING – 1990

Sl. No.	Themes
1.	Location map
2.	Relief map/Topography/Physical features
3.	Drainage map – showing the watersheds
4.	Soil map
5.	Monthly average rainfall for different blocks and the number of rainy days
6.	Land use map giving the percentage of area under different uses to the total geographical area for different blocks as well as for the district as whole
7.	Cropping pattern – Area under major crops in the block to be shown in pie-diagram as percentage to the total cropped area
8.	Drought prone areas
9.	Irrigation facilities — existing and proposed
10.	SC/ST population as percentage to the total population shown in bar diagram in the map showing the tribal and non-tribal areas
11.	Agricultural infrastructural facilities — existing and proposed
12.	Veterinary facilities etc. — existing and proposed
13.	Cattle population block-wise in a map alongwith milk route and chilling plants
14.	Ground water potentiality alongwith feasible number of wells
15.	Drinking water-problem village-category: A, B, C
16.	Industrial estates, factories etc.
17.	Waste land and forest map
18.	Places of religious and historical importance
19.	Transport network — existing and proposed

20. Transmission lines, transformers along with the capacity, villages to be electrified etc.
21. Post and telegraphic offices
22. Financial institutions and service area
23. Educational institutions — existing and proposed
24. Health facilities — existing and proposed
25. Central villages and service area

APPENDIX — VIII

LIST OF THEMES BASED ON REVIEW OF GUIDE-LINES FOR DISTRICT LEVEL PLANNING

Sl. No.	Themes
1.	Location
2.	Relief
3.	Monthly rainfall
4.	Seasonal rainfall
5.	No. of rainy days
6.	Drainage
7.	Watersheds
8.	Soil types
9.	Depth of underground water table
10.	Ground water potentiality
11.	Natural vegetation
12.	Forests
13.	Land use
14.	Distribution of wastelands
15.	Cropping pattern
16.	Percentage of area under forest
17.	Percentage of barren and uncultivable land
18.	Percentage of cultivable waste land
19.	Percentage of area under fallows other than current fallows
20.	Percentage of area under current fallows
21.	Percentage of area under permanent pasture and grazing lands
22.	Percentage of area under non-agricultural use
23.	Percentage of area under miscellaneous tree crops
24.	Percentage of net sown area
25.	Sources of irrigation
26.	Location and command of irrigation projects
27.	Cattle population
28.	Milk chilling plants
29.	Milk routes

30. Rural-urban composition
31. Age-sex composition
32. Distribution of population
33. Population size-wise distribution of settlements
34. Concentration Scheduled Castes population
35. Concentration Scheduled Tribes population
36. Concentration of land-less agricultural labourers
37. Location of education facilities
38. Location of health facilities
39. Location of veterinary facilities
40. Location of hats, bazaars and fairs
41. Agri. Infrastructural facilities: Storage godown
42. Seed and fertilizer distribution
43. Location of banks
44. Location of credit facilities
45. Location of service area of banks
46. Location of transportation network
47. Location of post & telegraph offices
48. Location of power distribution work (with transformers)
49. Location of villages to be electrified
50. Drinking water facilities — Sources and quality of water
51. Drinking water problem villages
52. Location of mines
53. Location of flour mills
54. Location of rice mills
55. Location of oil expellers
56. Location of other industrial units
57. Location of industrial estates
58. Location of places of religious and historical importance
59. Problem areas : Area liable to flooding
60. Waterlogging
61. Soil erosion
62. Alkalinity
63. Recurrent drought
64. Levels of development

APPENDIX — IX

THEMES SELECTED BY THE MAP USER FROM THE LIST OF MAPS/MAP THEMES SUPPLIED BY THE RESEARCHER

Sl. No.	Themes	
1.	Location	
2.	Administration	
3.	Relief	
4.	Slope	
5.	Drainage	
6.	Watershed boundary	
7.	Geology	
8.	Soil	
9.	Chemical Properties of Soil:	Sodium
10.		Phosphorous
11.		Potassium
12.	Annual rainfall	
13.	Seasonal rainfall	
14.	Annual temperature	
15.	Natural vegetation	
16.	Forest	
17.	Depth of water table	
18.	Ground water quality	
19.	Ground water potential	
20.	Sources of irrigation	
21.	Extent of irrigation	
22.	Area irrigated by well	
23.	Area irrigated by tank	
24.	Area irrigated by canal	
25.	Area irrigated by other sources	
26.	Land use	
27.	Percentage of barren and uncultivated land	
28.	Percentage of cultivable waste land	
29.	Percentage of area under fallows other than current fallows	

30. Percentage of area under current fallows
31. Percentage of area under permanent pastures and grazing lands
32. Percentage of land under miscellaneous tree crops
33. Percentage of net sown area
34. Area under kharif crops
35. Area under rabi crops
36. Area under Paddy
37. Area under Groundnut
38. Area under Maize
39. Area under Commercial crops
40. Man-land ratio
41. Intensity of cropping
42. Distribution of wastelands
43. Land holdings of small and marginal farmers
44. Land holdings of tenant cultivators: Area
45. Land holdings of tenant cultivators: Number
46. Tractor and tractor operated implements
47. Bullock and manually operated implements
48. Oil engines, electrical pumpsets and sprayers
49. Sugarcane crushers, oil ghannies and thrusters
50. Distribution of livestock
51. Milch animal
52. Markets for livestock
53. Density of population
54. Literates
55. Male literates
56. Birth, death and infant mortality rates
57. Total Scheduled Castes population
58. Total Scheduled Tribes population
59. Occupational structure
60. Percentage of agricultural labourers
61. Distribution of urban and rural settlements
62. Supply of drinking water — drinking water problem villages
63. Power grid with transformers
64. Villages to be electrified
65. Large and medium scale industries: Workers and capital invested
66. Distribution of small scale industries

67. Agro-based industries:
    No. of units and workers
68. Capital invested and value of output
69. Agri. implements manufacturing industries:
    No. of units and workers
70. Capital invested and value of output
71. Wood based industries:
    No. of units and workers
72. Capital invested and value of output
73. Rubber & leather industries:
    No. of units and workers
74. Capital invested and value of output
75. Chemical industries:
    No. of units and workers
76. Capital invested and value of output
77. Ceramic and building material industries:
    No. of units and workers
78. Capital invested and value of output
79. Engineering industries:
    No. of units and workers
80. Capital invested and value of output
81. Electrical & allied industries:
    No. of units and workers
82. Capital invested and value of output
83. Printing, binding and allied industries:
    No. of units and workers
84. Capital invested and value of output
85. Misc. Industries:
    No. of units and workers
86. Capital invested and value of output
87. Roads and railways
88. Inland waterways
89. Airways and sea routes
90. Centres of education
91. School education districts
92. Banks
93. Service area of banks
94. Telegraph grid
95. Post and telegraph offices
96. Hospitals

97. Primary health centres
98. Veterinary facilities
99. Tourism
100. Problem areas: Areas liable to flooding

BIBLIOGRAPHY

Atlases

Administrative Atlas of Andhra Pradesh, Census of India, New Delhi, 1977.

Administrative Atlas of Haryana, Census of India, New Delhi, 1972.

Administrative Atlas of Rajasthan, Census of India, New Delhi, 1976.

Administrative Atlas of Tamil Nadu, Census of India, New Delhi, 1974.

Agricultural Atlas of India, Directorate of Land Records, Madhya Pradesh, Gwalior, 1959.

Agro-Climatic Atlas of India, Indian Meteorological Department, Pune, 1978.

Ambasthya, B.P., *James Rennell's Memoir of a Map of Hindustan and the Bengal Atlas*, N.V. Publications, Patna, 1975.

Atlas of Agricultural Resources of India, National Atlas and Thematic Mapping Organisation, Calcutta, 1980.

Atlas of Forest Resources of India, National Atlas and Thematic Mapping Organisation, Calcutta, 1976.

Atlas of Indian Lac, Indian Lac Cess Committee, Ranchi, 1958.

Burdwan District Atlas, Department of Development and Planning, Calcutta Metropolitan Planning Organisation, Calcutta, 1969.

Census Atlas of India, Census of India, New Delhi, 1970.

Chatterjee, S.P., *Planning Atlas of Damodar Valley Region,* Damodar Valley Corporation, Calcutta, 1969.

Climatological Atlas of India, Indian Meteorological Department, Pune, 1971.

Coconut Atlas of India, Indian Central Coconut Committee, Ernakulam, 1959.

Diet Atlas of India, National Institute of Nutrition, Indian Council for Medical Research, Hyderabad, 1959.

Duggar, S.L., *Agricultural Atlas of Punjab,* Agricultural University, Hissar, 1966.

Economic and Statistical Atlas of Madhya Pradesh, Directorate of Economics and Statistics, Bhopal, 1958.

Economic Atlas of Madras State, National Council for Applied Economic Research, New Delhi, 1962.

Government of Andhra, *Statistical Atlas of Andhra State,* Bureau of Economics and Statistics, Hyderabad, 1959.

Government of India, *Atlas of Northern Frontier of India,* Ministry of External Affairs, New Delhi, 1960.

Government of India, *Indian Agricultural Atlas*, Ministry of Food & Agriculture, New Delhi, 1958.

Government of India, *Indian Agricultural Atlas,* Department of Agriculture, New Delhi, 1975.

Government of India, *National Atlas of India* (Hindi Edition), Ministry of Education and Scientific Research, Calcutta, 1957.

Government of Kerala, *Kerala in Maps*, Bureau of Economics and Statistics, Trivandrum, 1978.

Government of Mysore, *Mysore State in Maps,* Department of Statistics, Bangalore, 1958.

Government of Orissa, *Statistical Atlas of Orissa,* Bureau of Statistics and Economics, Bhubaneswar, 1959.

Government of Pondicherry, *Statistical Atlas of Union Territory,* Bureau of Statistics and Valuation, Pondicherry, 1976.

Government of Rajasthan, *Statistical Atlas of Rajasthan*, Directorate of Economics and Statistics, Jaipur, 1959.

Indian Census Centenary Atlas, Census of India, New Delhi, 1979.

Indian Cotton Atlas, Central Cotton Committee, Bombay, 1959.

Indian Jute Atlas, Indian Central Jute Committee, Calcutta, 1959.

Indian Literacy Atlas, Central Institute of Indian Languages, Mysore, 1978.

Indian Oilseeds Atlas, Indian Central Oilseeds Committee, Hyderabad, 1958.

Indian Tobacco Atlas, Indian Central Tobacco Committee, Madras, 1959.

Irrigation Atlas of India, National Atlas and Thematic Mapping Organisation, Calcutta, 1972.

Learmonth, L.T.A., and Bhat, L.S. *Mysore State: An Atlas of Resources,* Calcutta, 1961.

Manzoor, Alam., *Planning Atlas of Andhra Pradesh,* Osmania University, Hyderabad, 1974.

Mohammed, A., *Planning Atlas of Medak District,* Centre for Economic and Social Studies, Hyderabad, 1988.

National Atlas of India, National Atlas and Thematic Mapping Organization, Calcutta, 1982.

Nutrition Atlas of India, National Institute of Nutrition, Indian Council for Medical Research, Hyderabad, 1959.

Rainfall Atlas of India, Indian Meteorological Department, Pune, 1971.

Ramesh, A., and Tiwari, P.S., *Basic Resource Atlas of Tamil Nadu,* University of Madras, Madras, 1983.

Resource Atlas of Kerala, Centre for Earth Science Studies, Trivandrum, 1984.

Sen, A.K., *Agricultural Atlas of Rajasthan*, Indian Council of Agricultural Research, New Delhi, 1970.

Singh, J., *Agricultural Atlas of India,* Vishal Publications, Kurukshetra, 1974.

Social and Economic Atlas of India, Oxford University Press, Madras, 1987.

Tiwari, P.K., *Agricultural Atlas of Uttar Pradesh*, G.B. Pant University of Agriculture and Technology, Pantnagar, 1973.

Tourist Atlas of India, National Atlas and Thematic Mapping Organisation, Calcutta, 1974.

Books

Bhat, L.S. *et. al., Micro Level Planning — A Case Study of Karnal Area, Haryana, India,* K.B. Publications, New Delhi, 1976.

Goel, S., *India within the Ganges,* Jaya Printers, New Delhi, 1983.

Harmensen, T., *Spatial Organisation and Economic Development,* Institute of Development Studies, University of Mysore, Mysore, 1971.

Maithani, B.P., *Spatial Analysis in Micro Level Planning*, Omsons Publications, Guwahati, 1986.

Mandle, S., *District Planning in India,* Indian Institute of Public Administration, New Delhi, 1977.

Marrison, J.L., The Science of Cartography and its Essential Process, *International Year Book of Cartography*, XVI, 1976.

Misra, R.P., Sundaram, K.V., and Rao, V.L.S.P., *Regional Development Planning in India — A New Strategy,* Vikas Publishing Pvt. Ltd., New Delhi, 1974.

Misra, R.P., (Ed), *Contribution to Indian Geography-I, Concepts and Approaches,* Heritage Publications, New Delhi, 1980.

Naidu, K.M., (Ed), *Area Planning for Regional Development,* Inter-India Publications, New Delhi, 1984.

National Informatics Centre, *District Information System of National Informatics Centre, DISNIC*, DISNIC Handbook, Government of India, Planning Commission, New Delhi, 1992.

Nikishov, M.L., and Preobrazhensky, "The problems of the Unification of the Contents and Conventional Signs Standardization on Maps", *International Year Book of Cartography*, Vol. II, 1971.

Otto, C. Strossel., *Standard Printing Screen System,* USAF Aeronautical Chart and Information Center, St.Louis, 1972.

Prasad, K., *Planning at the Grass Roots*, Sterling Publications Pvt. Ltd., New Delhi, 1988.

Raisz, E., *Principles of Cartography*, McGraw-Hill, New York, 1962.

Reddy, V.V., *Multi-Level Planning in India,* Vikas Publishing Pvt. Ltd., New Delhi, 1979.

Robinson, A.K., and Sale, R.D., *Elements of Cartography,* Third Edition, John Wiley & Sons, New York, 1969.

Robinson, A.K., Sale, R.D., and Merrison, J., *Elements of Cartography,* Fourth Edition, John Wiley & Sons, New York, 1978.

Sandor Rado, and Tibor Dudar., "Some Problems of Standardization of Transportation Map Symbols in Thematic Mapping", *International Year Book of Cartography,* Vol. II, 1971.

Singh, R.L., and Singh R., *Map Work and Practical Geography,* Central Book Depot, Allahabad, 1973.

Stephane, de, Brommer., "Proposal for Standardisation of Symbols on Thematic Maps", *International Year Book of Cartography,* Vol. 6, 1966.

Taylor, F.A., *Colour Technology,* Oxford University Press, London, 1962.

Tiwari, P.C., *Regional Development and Planning in India*, Criterion Publications, New Delhi, 1988.

Vyas V.S., Anil Bhat and Shah S.M., *Decentralised Planning in India,* Oxford and IBH Publishing Co., New Delhi, 1985.

Encyclopaedia/Dictionary

Chambers 20th Century Dictionary, Allied Publishers, New Delhi, 1986.

David, H., Baily., *et. al.,* (Eds), *Encyclopaedia Britanica,* McGraw-Hill Book Co., New York, 1979.

Fried, H. Herman., *et.al.,* (Eds), *The University Desk Encyclopaedia,* E.P. Dalton Inc, New York, 1977.

Steiner Verlag, *Multilingual Dictionary of Technical Terms in Cartography,* Wicsbader, FRA, 1973.

Journals

Ahmed, M., "Search for more effective methods in Population Mapping", *Indian Cartographer,* Vol. V, 1985.

Alan A. Delucia., "The Effect of Shaded Relief on Map Information", *Cartographic Journal,* Vol. 9, No.1, 1972.

Alan A. Delucia and Donald, W. Hillor., "Natural Legend Design for Thematic Maps", *Cartographic Journal,* Vol. 19, No.1, 1982.

Anatolji, M.Kamkov., "The International Language of Geographical Maps", *International Year Book of Cartography,* Vol. 19, 1971.

Arunachalam B., "Status of Cartography Education in India and Measures for its Quantitative Improvement", *Indian Cartographer,* Vol. VII, 1987.

Barbara, S. Barty., "Experimental Use of Search Task in an Analysis of Type Legibility in Cartography", *Cartographic Journal* Vol. 7, No. 2, 1970.

Bartholomew, C. John., and Kinniburgh A.C. Ian., "The Factors of Awareness" *Cartographic Journal,* Vol. 10, No.1, 1973.

Berger, R.V., "An International Study of Children's Perception and Understanding of Type Used on Atlas Maps", *Cartographic Journal,* Vol. 19, No. 2, 1982.

Bhatia A.K and Nagarajan B., "Computer Assisted Cartography — Its Application in Micro-Level Planning and Regional Development", *Indian Cartographer,* Vol. VIII, 1987.

Das Gupta S.P., "Trends in Thematic Cartography", *Indian Cartographer,* Vol. I, 1980.

——, "Growth and Development of Cartography in India", *Indian Cartographer,* Vol. IV, 1982.

David J. Cuff., "Shading on Choropleth Maps: Some Suspicions Confirmed", *Proceedings of the Association of American Geographers,* Vol. 5, 1973.

——, "Colour on Temperature Maps", *Cartographic Journal,* Vol. 10, No.1, 1973.

Ganesh, P.V., "District Maps for Developmental Plans upto Block Level", *Indian Cartographer,* Vol. VIII, 1988.

George F. Jenks., and Duane S. Knos, "The Use of Shading Patterns in Graded Series", *Annals of the Association of American Geo-*

grapher, Vol. 51, No. 3, 1961.

___ , "Error on Choropleth Maps: Definition, Measurement, Reduction", *Annals of the Association of American Geographer,* Vol. 6, No. 2, 1971.

Goel Susan., "Indigenous Cartography in India" *Indian Cartographer,* Vol. V, 1985.

Goesta, Ekman and Kenneth, Junge; "Psychological Relation in Visual Perception of Length, Area and Volume", *Scandinavian Journal of Psycology,* Vol. 2, No.1, 1961.

Grant Head, C., "Land Water Differentiation in Black and White Cartography", *Canadian Cartographer,* Vol. 9, No. 1, 1972.

Hans, J., Maihafer., "The Utility of the Circle as an Effective Cartographic Symbol", *Canadian Cartographer,*Vol. 6, 1969.

___ , "The Visual Perception of the Circle in Thematic Maps/Experimental Results", *Canadian Cartographer,* Vol. 10, No.1, 1973.

___ , "The Relative Effectiveness of Some Common Graduated Point Symbols in the Presentation of Quantitative Data", *Canadian Cartographer,* Vol. 8, No. 2, 1971.

Jean, Clande, Muller., "Association in Choropleth Map Comparison", *Annals of the Association of the American Geographer,* Vol. 65, No. 3, 1975.

Jeffrey, C. Pattom., and Paul, V. Crawfold., "The Perception of Hypsometric Colours", *Cartographic Journal,* Vol. 14, No. 2, 1977.

John, I.Clarke., "Statistical Map Reading", *Geography,* Vol. 44, No. 2, 1959.

Joshi, V.C., 'Village Maps for Rural Development', *Indian Cartographer,* Vol. VIII, 1988.

Kand, T. Chang., "Visual Aspects of Class Intervals in Choroplethic Mapping", *Cartographic Journal,* Vol. 15, No. 1, 1978.

Kolonocny, A., "Cartographic Information- A Fundamental Concept and Term in Modern Cartography", *Cartographic Journal,* Vol. 6, No. 1, 1969.

Lech, Ratjaski., "The Methodical Basis of the Standardization of Signs on Economic Maps", *International Year Book of Carto-graphy,* Vol. II, 1971.

Maithani, B.P., Tiwari, P.C. and Pandey, D.C. "Spatio-Functional Organisation for Integrated Rural Development Planning", *The Deccan Geographer,* Secundarabad, 1982.

Mark, S. Monmonier., "Class Intervals to Enhance the Visual Correlation of Choropleth Maps", *Canadian Cartographer,* Vol. 12, No. 2,

1975.

Mark Shurtleft, and R. Edward Geiselman, "A Human Performance based Evaluation of Topographic Maps and Map Symbols with Novice Map Users", *Cartographic Journal,* Vol. 23, No. 1, 1986.

Nambison, I.M.G. and Mahendra, H.C., "Atlas Cartography", *Indian Cartographer,* Vol. IV, 1982.

Nikishov, M.L., and Preobrazhensky. "The Problems of the Unification of the Contents and Conventional Signs Standardization on Maps", *International Year Book of Cartography,* Vol. II, 1971.

Patrick, Sorrell., "Map Design – With the Young in Mind", *Cartographic Journal,* Vol. II, No. 2, 1974.

Paul, V. Crawford., "Perception of Graduated Squares as Cartographic Symbols", *Cartographic Journal,* Vol.10, No. 2, 1973.

Richard, Philips, De Lucia Alan, and Skelton Nicholas, "Some Objective Tests of the Legibility of Relief Maps", *Cartographic Journal,* Vol. 12, No. 1, 1975.

Richard, J.Philips and Liza Noyes., "An Investigation of Visual Clutter in the Topographic Base of a Geological Map", *Cartographic Journal,* Vol.19, No. 2, 1982.

Robert, L.Williams., "Map Symbols: Equal Appearing Intervals for Printed Screens", *Annals of the Association of American Geographer,* Vol. 48, No. 2, 1958.

Rose J. Mackey., "Experiments with Some Symbols and Map Projections", *Annals of the Association of the American Geographer,* Vol. 44, No. 2, 1954.

Sharma L.P., "Modern Cartographic Education Trends", *Indian Cartographer,* Vol. III, 1987.

Sheppard, D., and Adam, J.M., "A Survey of Driver's Opinion on Maps for Route Finding", *Cartographic Journal,* Vol. 8, No. 2, 1971.

Singh, A.K., "Decentralised Planning under the Five Year Plans: Emerging Perspectives", *Journal of Rural Development,* Vol. VIII, No. 3, 1989.

___, "Decentralised Planning in U.P. — A Critical Appraisal", *Indian Journal of Regional Science,* Vol. XIX, No.1, 1987.

Singh, Mohinder., "Introduction to History of Indian Cartography", *Indian Cartographer,* Vol. III, 1981.

Somaiah, T.N., "Cartography in the Development of District Level Planning", *Indian Cartographer,* Vol. V, 1985.

Taylor, R.M., "An Application of Subjective Scaling Techniques to Map Evaluation", *Cartographic Journal,* Vol. 11, No. 2, 1974.

News Papers / Articles

Banerjee, K.Utpal., "Information System in District Planning", *The Hindu,* June 1, 1989.

___ , "Information Gathering from District Level" *The Hindu,* June 8, 1989.

___ , "Key Consideration in Data Collection, *The Hindu,* June 22, 1989.

Gurusamy, T.K., Unni, C.K.P., and Chandwaskar, A.M., "Cartographic output by Digitization on Micro Processor", *A Paper,* IV Annual Conference of Indian National Cartographic Association, Madras, 1982.

Hellen Wallis., Cartographic Innovation: An Historical Perspective Centennial Convention, *A Paper,* Congress Du Centenaire, Ottawa, 1982.

Government of India, Ministry of Environment & Forests, INDO-US Workshop on Geographic Information System, *Background papers,* New Delhi, 1989.

Moni, M., and Vijayaditya N., DISNIC -- A NICNET based District Government Information Programme in India, *A Paper,* Indian Computing Congress, New Delhi, 1991.

Satchidanand, N.N., India Entering Information on Top Era, *The Hindu,* July 4, 1990.

Reports

Government of India, Administrative Reforms Commission, *Report of the Study Team on the Machinery for Planning*, New Delhi, 1977.

Government of India, *Report of the Committee to Review the Existing Administrative Arrangements for Rural Development and Poverty Alleviation Programme,* Ministry of Agriculture, Department of Rural Development, New Delhi, 1985.

Government of India, *Report of the Committee on Panchayat Raj Institutions,* Ministry of Agriculture and Irrigation, New Delhi, 1978.

Government of India, *I Five Year Plan,* Planning Commission, New Delhi, 1952.

Government of India, *II Five Year Plan,* Planning Commission, New Delhi, 1956.

Government of India, *III Five Year Plan*, Planning Commission, New Delhi, 1961.

Government of India, *VII Five Year Plan,* Planning Commission, Vol. II, New Delhi, 1985.

Government of India, *Report of the Working Group on Block Level Planning,* Planning Commission, New Delhi, 1978.

Government of India, *Report of the Working Group on District Planning,* Vol. I, Planning Commission, New Delhi, 1984.

Government of India, *Report of the Working Group on District Planning,* Vol. II, Planning Commission, New Delhi, 1984.

Government of India, *Report of the Study Team on Community Development and National Extension Service,* New Delhi, 1983.

Government of India, Performance Budget 1991-92, Ministry of Planning, Planning Commission, New Delhi, 1991.

Government of India, Performance Budget 1992-93, Government of India, Ministry of Planning, Planning Commission, New Delhi, 1992.

Gurusamy, T.K., *Annual Report of INCA Commission for School Atlas*, Fourth National Cartographic Conference, Madras, 1982.

James, J. Flannery., "The Graduated Circle: A Description, Analysis and Evaluation of a Quantitative Map Symbol", Unpublished *Doctoral Dissertation*, University of Wisconsin, Wisconsin, 1965.

McGrath, G., and Castner, H.W., "The Design of a Tourist Map Series based on the National Topographic 1:250000 Map Series", First *Report* Under Contract. OISU 23244, 5-5596 Department of Energy, Mines and Resources, Surveys and Mapping Branch, Topographical Survey Directorate.

National Informatics Centre, General Information Service Manual, Government of India, Planning Commission, New Delhi, 1991.

Thaha, M., *et al.*, *Guidelines for District Planning,* National Institute of Rural Development, Hyderabad, 1990.

United Nations, *Report of the Committee of Experts on Cartography,* 'Modern Cartography – Base Maps for World Needs', U.N. Publications, 1949.

INDEX